AIR SPARGING FOR SITE REMEDIATION

Edited by

Robert E. Hinchee
Battelle, Columbus, Ohio

LEWIS PUBLISHERS
Boca Raton   Ann Arbor   London   Tokyo

Library of Congress Cataloging-in-Publication Data

Catalog record is available from the Library of Congress.

International Standard Book Number 1-56670-084-1

CONTENTS

Articles

Technical Notes

FOREWORD

The need or demand for soil and groundwater remediation technology has far outstripped the development of remedial technology. Over the past 10 years hundreds of thousands of regulatory actions at sites throughout the world have required soil or groundwater remediation, often to levels that have never been achieved, or at least have not been achieved in a clearly documented, reproducible fashion. Particularly with regard to remediation of aquifers contaminated with hydrophobic organics, few field data exist to support most claims of remediation. The point is not that sites are not being remediated, but that the state of the practice of remediation remains very much a "black box."

Air sparging is one of those technologies that falls into this black box. Based on field experience we know with certainty that air sparging can remove and degrade contaminants. The mechanisms for the removal, however, are not clearly understood. As a result, the state of the art of design relies very much on the professional judgment of individual designers. That professional judgment typically is based on experience. No widely accepted rational design approach has yet emerged. The other problem arising from a lack of understanding of the mechanisms by which air sparging works is the inability to predict achievable cleanup levels and project cleanup times. The most reliable projections currently can be based only on experience, and experience with well-documented applications is far too limited.

In evaluating any site remediation technology, comparisons to competing technologies are necessary. This is where air sparging looks favorable. To varying extents, the concerns raised above apply to all in situ technologies. The problems are that in situ remediation is very difficult to adequately monitor and is very expensive to study with sufficient analysis to provide a good mass balance at even the simplest of sites.

Groundwater pump-and-treat currently is the most commonly applied technology for aquifer remediation and, in many cases, it may hydraulically contain a plume, preventing further migration. We now know that, to remediate a contaminated aquifer, groundwater pump-and-treat is at best a very slow and limited approach. In situ bioremediation is another aquifer treatment technology. In practice, it has been used primarily for remediation of petroleum hydrocarbons. The primary limiting factor for in situ bioremediation usually is oxygen. In addition to stimulating volatilization, air sparging appears to offer a more efficient means of oxygen delivery than other methods now in practice. Although many other in situ hydrocarbon remediation technologies are in various stages of development, none offers the potential for application to aquifer remediation that air sparging does.

The intent of this book is to begin prying open the black box of aquifer remediation and to offer some insight into the process.

This book and four other volumes represent papers arising from the Second International Symposium on In Situ and On-Site Bioreclamation held in San Diego, California, in April 1993. The other four books are entitled *Bioremediation of Chlorinated and Polycyclic Aromatic Hydrocarbons, Hydrocarbon Bioremediation, Applied Biotechnology for Site Remediation,* and *Emerging Technology for Bioremediation of Metals.*

The symposium was attended by more than 1,100 people. More than 300 presentations were made, and all presenting authors were asked to submit manuscripts. Following a peer review process, 190 papers are being published. The editors believe that these volumes represent the most complete, up-to-date works describing both the state of the art and the practice of bioremediation.

With the exception of the first paper, authored by the editor, the papers in this book were part of the Second International Symposium. That symposium was sponsored by Battelle Memorial Institute and was cosponsored by:

Bruce Bauman, *American Petroleum Institute*

Christian Bocard, *Institut Français du Pétrole*

Rob Booth, *Environment Canada, Wastewater Technology Centre*

D. B. Chan, *U.S. Naval Civil Engineering Laboratory*

Soon H. Cho, *Ajou University, Korea*

Kate Devine, *Biotreatment News*

Volker Franzius, *Umweltbundesamt, Germany*

Giancarlo Gabetto, *Castalia, Italy*

O. Kanzaki, *Mitsubishi Corporation, Japan*

Dottie LaFerney, *Stevens Publishing Corporation*

Massimo Martinelli, *ENEA, Italy*

Mr. Minoru Nishimura, *The Japan Research Institute, Ltd.*

Chongrak Polprasert, *Asian Institute of Technology, Thailand*

Lewis Semprini, *Oregon State University*

John Skinner, *U.S. Environmental Protection Agency*

Esther Soczo, *National Institute of Public Health and Environmental Protection, The Netherlands*

In addition, the following individuals assisted as session chairs, presented invited papers, and helped to ensure diverse representation and quality:

Bruce Alleman, *Battelle Columbus*

Christian Bocard, *Institut Français du Pétrole*

Rob Booth, *Environment Canada, Wastewater Technology Center*

Fred Brockman, *Battelle Pacific Northwest Laboratories*

Tom Brouns, *Battelle Pacific Northwest Laboratories*

Soon Cho, *Ajou University, Korea*

M. Yavuz Corapcioglu, *Texas A&M University*

Jim Fredrickson, *Battelle Pacific Northwest Laboratories*

Giancarlo Gabetto, *Area Commerciale Castalia, Italy*

Terry Hazen, *Westinghouse Savannah River Laboratory*

Ron Hoeppel, *U.S. Naval Civil Engineering Laboratory*

Yacov Kanfi, *Israel Ministry of Agriculture*

Richard Lamar, *U.S. Department of Agriculture*

Andrea Leeson, *Battelle Columbus*

Carol Litchfield, *Keystone Environmental Resources, Inc.*

Perry McCarty, *Stanford University*

Jeff Means, *Battelle Columbus*

Blaine Metting, *Battelle Pacific Northwest Laboratories*

Ross Miller, *U.S. Air Force*

Minoru Nishimura, *Japan Research Institute*

Robert F. Olfenbuttel, *Battelle Columbus*

Say Kee Ong, *Polytechnic University, New York*

Augusto Porta, *Battelle Europe*

Roger Prince, *Exxon Research and Engineering Co.*

Parmely "Hap" Pritchard, *U.S. Environmental Protection Agency*

Jim Reisinger, *Integrated Science & Technology, Inc.*

Greg Sayles, *U.S. Environmental Protection Agency*

Lewis Semprini, *Oregon State University*

Ron Sims, *Utah State University*

Marina Skumanich, *Battelle Seattle Research Center*

Jim Spain, *U.S. Air Force*

Herb Ward, *Rice University*

Peter Werner, *University of Karlsruhe, Germany*

John Wilson, *U.S. Environmental Protection Agency*

Jim Wolfram, *Montana State University*

The papers in this book have been through a peer review process, and the assistance of the peer reviewers is recognized. This typically thankless job is essential to technical publication. The following people peer-reviewed papers for this volume:

Richelle M. Allen-King,
University of Waterloo
James F. Barker, *University of
Waterloo*
Tad Beard, *Battelle Columbus*
Jason A. Caplan, *ESE Biosciences*
Edward Coleman, *MK
Environmental*
Dave DiPauli, *Oak Ridge National
Laboratory*
Perry Hubbard, *Integrated
Science & Technology, Inc.*
Yacov Kanfi, *Israel Ministry of
Agriculture*
Chih-Ming Kao, *North Carolina
State University*

Paul R. Kurisko, *Envirogen, Inc.*
Perry L. McCarty, *Stanford
University*
Gloria McCleary, *EA Engineering*
Linda McConnell, *Logistics
Management Institute*
David B. McWhorter, *Colorado
State University*
Ralph E. Moon, *Geraghty &
Miller, Inc.*
Richard E. Perkins, *DuPont
Environmental Biotechnology
Program*
Thomas J. Simpkin, *CH2M HILL*
Rodney S. Skeen, *Battelle Pacific
Northwest Laboratories*

It must be pointed out that in more than one case papers were accepted and published despite peer reviewers not recommending publication of the papers based on what they believed to be unsubstantiated claims or questionable data interpretation. The reason for this was not that the judgment of the reviewers was questioned, but rather that these papers, despite technical awkwardness, are representative of the state of the practice. In reading the book many of these problems should become apparent. The technical community frequently is uncomfortable with processes that are not well understood, and the conservative approach is to question statements and conclusions based on limited data and understanding. This is not to say that these statements and conclusions are wrong, although some may prove to be, but that they simply are not verifiable given our current state of understanding of air sparging. The editor accepts full responsibility for the decision to publish these papers and thereby relieves the peer reviewers of any accountability.

The process of assembling this book has required a significant effort and the editor wishes to recognize some of the key contributors. Lynn Copley-Graves served as the text editor, reviewing every paper for readability and consistency. She also directed the layout of the book and production of the camera-ready copy. Loretta Bahn worked many long hours converting and processing files, and laying out the pages. Karl Nehring oversaw coordination of the book publication with the symposium and worked with the publisher to make everything happen. Gina Melaragno coordinated manuscript receipt and communications with the authors and peer reviewers.

None of the sponsoring or cosponsoring organizations or peer reviewers conducted a final review of the book or in any way has endorsed this book.

Rob Hinchee
June 1993

AIR SPARGING STATE OF THE ART

R. E. Hinchee

ABSTRACT ━━━━━━━━━━━━━━━━━━━━━━━━━━━━━━━━━━━

Air sparging by introducing air beneath the water table is being used to promote site remediation. The technology of air sparging involves two mechanisms working either alone or in tandem: volatilization and biodegradation. Most uses of air sparging mix the physical with the biological. Air sparging can be divided into two distinct technologies, in-well aeration and air injection into the aquifer. In-well aeration is the process of injecting gas, usually air, into a well, resulting in an in-well airlift pump effect. Air injection involves injection of air (or other gases) under pressure directly into saturated groundwater to provide oxygen for bio-remediation and/or strip or volatilize the contaminants out of the aquifer.

INTRODUCTION

The term *air sparging* is widely applied to the technology of intro-ducing air (or other gases) beneath the water table to promote site remediation. For contaminant treatment, air sparging relies on two basic mechanisms working either alone or in tandem: biodegradation and vola-tilization. In most cases air sparging is, therefore, a hybrid technology, being both physically and biologically based. Air sparging can be divided into two distinct technologies, in-well aeration and air injection into the aquifer.

In-well aeration is the process of injecting gas, usually air, into a well, resulting in an in-well airlift pump effect. The purpose is to decrease pressure in the deeper portions of the well and to increase pressure in the upper portion. The pressure changes result in groundwater flow into the well at the bottom and flow out in the upper portions of the screen. The airstream also may serve to strip volatiles and/or provide oxygen for biodegradation. The water flow results in a circulation pattern being

1-56670-084-1/94/$0.00 + $.50

established in the aquifer. This technology is most commonly referred to as "UVB" from the German *Unterdruck-Verdampfer-Brunnen*, translated as vacuum vaporizer well.

Air injection into the aquifer is a very different process. Air under pressure greater than that of the water depth is injected directly into groundwater-saturated aquifer materials. The objective is to force the air through contaminated aquifer materials to provide oxygen for bio-remediation and/or strip the contaminants out of the aquifer. Air injection is commonly referred to as *air sparging*.

IN-WELL AERATION

In-well aeration is the process of injecting into a well air that is not intended to enter the aquifer, except perhaps in a dissolved form. Even if bubbles were to leave the well it is unlikely that they could be trans-ported in most aquifer material. In a typical application, illustrated in Figure 1, air is injected into the bottom of a well. The air travels upward,

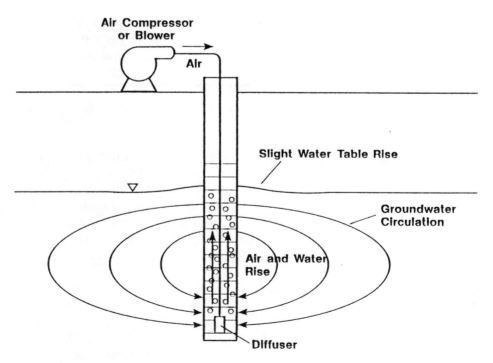

FIGURE 1. Typical in-well aeration application.

removing volatiles and aerating the water. This upward movement of air results in an airlift pump effect causing water to flow into the well from the deeper screened portion of the well and out of the well from the shallower screened portion. Depending on hydrogeologic conditions, a circulation cell is then established that treats and aerates the water as it passes through the well.

To the author's knowledge, the first documented use of in-well aeration air sparging was by Raymond's group (Raymond 1974, 1976) in their early in situ bioremediation experiments. The technology has been most widely developed and exploited by IEG Technologies Corporation in Germany, which developed the UVB process illustrated in Figure 2. The significant differences between the UVB technology and simple in-well aeration (Figure 1) are that the upper and lower screens are hydraulically separated and aeration is usually indicated by a vacuum. IEG has developed many variations on the basic UVB system shown in Figure 2. For example, in some application IEG uses in-well pumps to control water flow, has introduced in-well treatment units, and has developed specialized screens to reduce fouling problems. Gorelick and others working at Stanford have proposed further interesting modifications (MacDonald & Kitanidis 1993).

The principle upon which in-well aeration works is relatively straightforward. Air is used to strip and/or oxygenate water by establishing an "in-well pump-and-treat" system. The advantage of this approach compared to traditional pump-and-treat systems is that it avoids the necessity to lift water for aboveground treatment. That some kind of circulation pattern is established is clear, but determining its site-specific flow characteristics is another matter. Three significant limitations of conventional pump and treat, however, also appear to apply to in-well aeration:

1. Mass transfer within a heterogeneous geology from lower permeability portions of the aquifer to higher permeability flow paths.
2. Limited solubility of many contaminants.
3. Limited oxygen solubility for bioremediation.

Claims of increased efficiency due to sonic or other effects have been made, but to the author's knowledge these claims have not been proven. Although in-well aeration often may prove to be more efficient than conventional pump and treat, in-well aeration appears to have many of the same limitations.

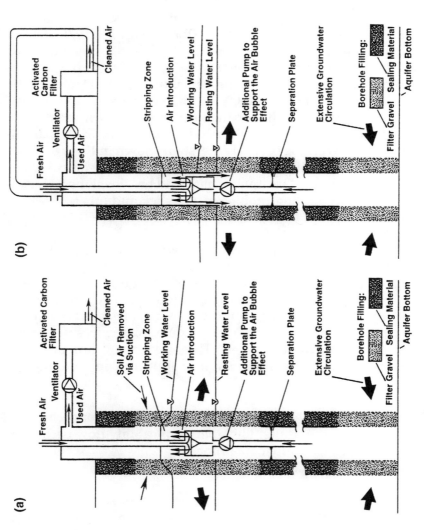

FIGURE 2. Vacuum vaporizer well (UVB): (a) with additional pump and separating plate and (b) with closed air circulation (from Herrling et al, 1991, reprinted with permission).

AIR INJECTION

Air injection into aquifer material works by different means than in-well aeration. Air injection occurs when a well screened below the water table, and hydraulically isolated from the vadose zone, is pressurized sufficiently to allow air flow into the aquifer. The air then migrates up through the aquifer material to the vadose zone where it may or may not be captured by extraction wells. Figure 3 illustrates a typical application of the process. Johnson et al. (1993) and others refer to this process as *in situ air sparging* or *IAS*. This term will be used to describe the process in this paper.

To the author's knowledge, IAS was first practiced in Europe (Hiller & Gudemann 1988), but the process is now being widely applied in the United States. A variety of configurations currently are in use. Billings et al. (1994) use nested clusters with relatively low air flow rates and limited radius of influence. Brown and Jasiulewicz (1992) use a higher air flow rate and appear to attain a greater radius of influence. Kampbell (1993) used an air-injection-only system (no soil gas extraction) at a gasoline-contaminated site and found that the volatilized hydrocarbons biodegraded before reaching the ground surface. Looney et al. (1991) utilized horizontal injection and extraction wells at the DOE site in Savannah River, Georgia.

IN SITU AIR SPARGING (IAS)

Process Mechanisms

The two mechanisms by which IAS works are volatilization and biodegradation. Volatilization evaporates and then extracts the contaminant. Biodegradation typically is stimulated by the introduction of oxygen. Other biodegradation processes are possible such as introduction of methane to stimulate cometabolism of chlorinated solvents. As more than 90% of the IAS bioremediation applications, of which the author is aware, inject air to stimulate biodegradation based on the process of oxygenation, this discussion is limited to oxygenation. Many of the observations, however, are valid for other air sparging bioremediation processes.

Air injected into aquifer materials migrates as a separate phase, typically in channels. It appears doubtful that bubbles form and migrate in any aquifers, except perhaps in highly permeable gravel aquifers. This is a very important observation. If bubbles do form and move, significant groundwater/air interaction will occur. The bubbles likely would induce

6

Air Sparging Air Sparging

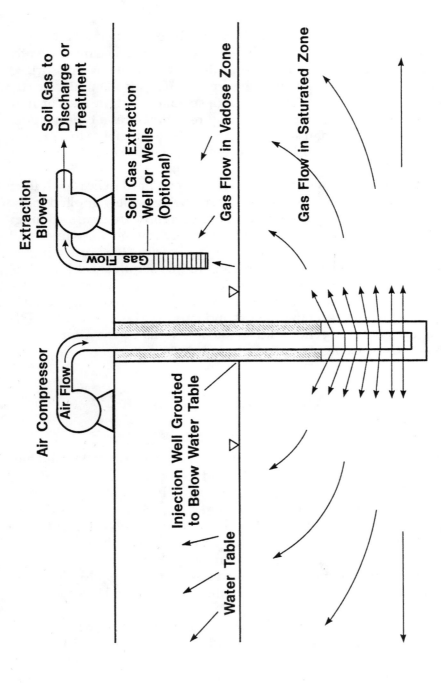

FIGURE 3. Typical in situ air sparging (IAS) application.

advective water flow, resulting in substantial air/water contact. If, however, bubbles do not form, as it appears they do not, air will flow in channels likely to have minimal interaction with water. Contaminated soil within those channels would be aerated; however, aquifer material not within these channels would be much less affected. Assuming channels do form, interaction between the channels and the surrounding water is important. Johnson (1994) has shown that stable air channels appear, and they do not seem to move with time or varying air flow rate. If these stable air flow channels do exist, the only opportunity for interaction may be diffusion. If a frictional interaction (i.e., drag) between the air and water occurs, water circulation will be induced. If and the extent to which this happens are not known; however, based on theoretical considerations, the effect appears to be minor and maybe insignificant. Varying the pressure within the air channels could result in changing channel diameters, thus inducing some water flow. However, this effect is speculative, certainly is unproven, and may not occur.

The mechanism of diffusion transfers volatile organics into the air channels and oxygen from the air channels into the aquifer. However, when diffusion works alone, the process is slow.

In saturated aquifer media the relative effects of volatilization and biodegradation, although analogous to volatilization vs. biodegradation in bioventing, likely will be quantitatively different. In vadose zone bioventing, the effects of aqueous solubility and Henry's law are less important than in the saturated zone. Miller et al. (1991) and Dupont et al. (1991) report that with vadose zone bioventing greater than 90% removal by biodegradation is possible. In the saturated zone, however, this may not be the case. Many hydrocarbon compounds have greater solubility and higher Henry's law constants than does oxygen, and as a result volatilization may become relatively more important. For example, oxygen has a solubility of approximately 9 mg/L and a Henry's law constant of approximately 1.5×10^{-3} atm $\cdot$ L/mmole (based on atmospheric O_2 concentrations) in contrast to benzene, which has approximate aqueous solubility and Henry's law constant of 1,780 mg/L and 4.4×10^{-3} atm $\cdot$ L/mmole, respectively.

IAS Monitoring

To date, most IAS has been monitored by observing dissolved hydrocarbon and oxygen concentrations in the wells, air pressure in the vadose zone, and the rise in the water table. The usefulness of these measurements in evaluating IAS is not clearly understood.

The rationale for measuring dissolved hydrocarbons or oxygen in monitoring wells is to attempt a direct measure of the success of the processes in removing volatiles and introducing oxygen. However, air bubbling up through monitoring wells frequently has been observed at IAS sites, leading to the misconception that air bubbles form in the aquifer. The presence of bubbles in a monitoring well does not indicate bubbles in the aquifer. Bubble formation is caused by air channels intercepting the well bore and allowing the air to rise vertically through the well. Figure 4 illustrates this phenomenon. The result is a well-aerated well. Some localized treatment of the aquifer surrounding the well may occur in a fashion similar to that described for in-well aeration, but most of the treatment would be limited to the well bore itself. As a result, measurements of dissolved hydrocarbons or oxygen in monitoring wells are most likely misleading, and it is possible that the presence of monitoring wells may reduce the IAS radius of influence and treatment efficiency. George Hoag's group at the University of Connecticut (George Hoag personal communication) is developing a probe specifically designed to overcome these problems and to discretely measure IAS impact on groundwater without impacting the air flow regime. To the author's knowledge, no such IAS groundwater monitoring approach is yet in general practice.

Monitoring air pressure measurements in the vadose zone does provide some indication of the influence of IAS on the vadose zone but does not appear to correlate with the effect on the underlying aquifer. The IAS impact on the vadose zone would be expected to be similar to that of soil venting or bioventing.

Many practitioners report a water table rise, and it does appear that the water table rises frequently in response to IAS application. This may be a transient phenomenon that occurs in response to air displacing the water (Johnson et al. 1990). Some practitioners explain this phenomenon as analogous to water surface rise in a well due to air injection, i.e., the principle by which an airlift pump works. This analogy does appear to be inappropriate. In the case of aeration in a free liquid, not in a saturated medium, air bubbles do form and, as they rise through the water, exert considerable viscous drag and cause turbulence as they move to the surface. This is very different from the phenomenon of air flow in discrete channels in a porous medium. Although a correlation may exist between the area in which the air is injected and this transient water table rise, it is not clear how this can be expected to correlate with the area of effective treatment.

One approach frequently used to illustrate IAS effectiveness and sometimes used as a pilot test is to initially extract air from the vadose zone

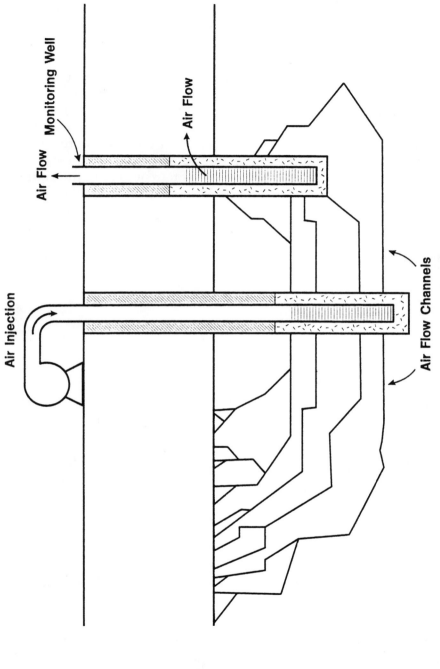

FIGURE 4. Cross section of IAS application illustrating air channeling to a monitoring well.

with no air injection. The rate of hydrocarbon removal by extraction and/or biodegradation is then measured. After a time air injection is initiated and changes in volatilization and/or biodegradation are observed. Figure 5 illustrates data of the kind typically observed. In most cases some increase in volatilization and/or biodegradation is observed after initiation of air injection. What is not known is the extent to which this effect is due to the removal of contaminants from the aquifer or to improved removal from the vadose zone. Does air injection actually improve long-term removal, or is it only a transient effect with no long-term net improvement? It is possible that the effect may be due to increased air flow in and treatment of the deep vadose zone rather than treatment in the aquifer.

SUMMARY AND CONCLUSION

In-well aeration and the UVB-type processes represent improvements on or new versions of conventional pump-and-treat technologies. In-well

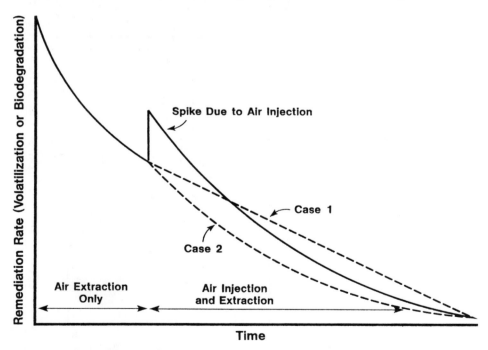

FIGURE 5. Typical removal curve utilized to illustrate IAS effectiveness. In case 1, no overall improvement in remediation is observed: in case 2, improvement is seen. To the author's knowledge, no demonstration data nor methods exist to determine which case is true.

aeration has the potential to be more cost effective and efficient than conventional pump-and-treat technology but will be subject to similar limitations. Much is to be learned about specific applications of in-well aeration, particularly regarding the questions of radius of influence and groundwater flow regime. The basic mechanisms controlling the process appear to be understood.

IAS may have the potential to significantly improve upon conventional groundwater treatment technologies. However, based on published studies to date and theoretical analyses, it also is possible that IAS may have a limited effect on aquifer contamination. The basic mechanisms controlling IAS are not well understood, and current monitoring practice does not appear to be adequate to quantitatively evaluate the process.

For many sites, no good conventional alternative to IAS exists. For example, many aquifers highly contaminated with hydrophobic hydro-carbons are not readily amenable to pump-and-treat technologies, or to any other conventional technologies. Neither lack of quantitative process data nor deficient understanding of the underlying mechanisms is limited to IAS. The state of the practice of aquifer remediation is such that regulatory pressures have driven and are driving widespread application of technologies for which adequate performance data do not exist. Aquifer remediation is still an inadequately understood process, and developing that understanding will necessarily be a slow and costly process. In the interim, regulatory pressures will require application of many poorly understood technologies. In such an environment IAS may well be the "least worst" alternative.

A significant problem in treating the vadose zone by soil venting or bioventing is inducing flow and mass transfer in the capillary fringe, i.e., the deep vadose zone immediately above the water table. It is entirely possible that air injection below the water table, i.e., IAS, will prove to be a more efficient means of inducing air flow in this region than the use of conventional vadose zone vent wells. Leeson et al. (1993) used air injection below the water table to improve air flow for bioventing.

IAS does present some risks. For example, some practitioners report contaminant spreading due to the water table rise, and there is an apparent risk of off-site vapor migration. On balance, however, at many sites these risks can be mitigated and IAS does not appear to pose greater risks than many other technologies. IAS should not be signi-ficantly more expensive than soil venting or bioventing and therefore should be economically attractive at many sites. If properly designed, IAS should do no harm and may enhance remediation. Future research should be aimed at clarifying and quantifying the effect of IAS on aquifer remediation.

REFERENCES

Billings, J. F., A. I. Cooley, and G. K. Billings. 1994. "Microbial and Carbon Dioxide Aspects of Operating Air-Sparging Sites." In R. E. Hinchee (Ed.), *Air Sparging*. Lewis Publishers, Ann Arbor, MI. pp. 112-119.

Brown, R. A., and F. Jasiulewicz. 1992. "Air Sparging Used to Cut Remediation Costs." *Pollution Engineering*, July. pp. 52-57.

Dupont, R. R., and W. J. Doucette. 1991. "Assessment of In Situ Bioremediation Potential and the Application of Bioventing at a Fuel-Contaminated Site." In R. E. Hinchee and R. F. Olfenbuttel (Eds.), *In Situ Bioreclamation*. Butterworth-Heinemann, Stoneham, MA. pp. 262-282.

Herrling, B., J. Stamm, and W. Buermann. 1991. "Hydraulic Circulation System for In Situ Bioreclamation and/or In Situ Remediation of Strippable Contamination." In R. E. Hinchee and R. F. Olfenbuttel (Eds.), *In Situ Bioreclamation*. Butterworth-Heinemann, Stoneham, MA. pp. 173-195.

Hiller, D., and H. Gudemann. 1988. "In Situ Remediation of VOC Conaminated Soil and Groundwater by Vapor Extraction and Groundwater Aeration." *Haztech International '88*, pp. 2A 90-110. Cleveland, OH, September.

Johnson, P. C., C. C. Stanley, M. W. Kemblowski, D. L. Byers, and J. D. Colthart, 1990. "A Practical Approach to the Design, Operation, and Monitoring of In Situ Soil-Venting Systems." *Groundwater Monitoring Review, 10*(2):159-178.

Johnson, R. L. 1994. "Enhancing Biodregradation with In Situ Air Sparging: A Conceptual Model." In R. E. Hinchee (Ed.), *Air Sparging*. Lewis Publishers, Ann Arbor, MI. pp. 14-22.

Johnson, R. L., D. B. McWhorter, P. C. Johnson, R. E. Hinchee, and I. Goodman. 1993 (in press). "An Overview of Air Sparging." *Groundwater Monitoring and Remediation*.

Kampbell, D. 1993. "U.S. EPA Air Sparging Demonstration at Traverse City, Michigan." In Environmental Restoration Symposium. Sponsored by the U.S. Air Force Center for Environmental Excellence, Brooks AFB, TX.

Leeson, A., R. E. Hinchee, J. Kittel, G. Sayles, C. Vogel, and R. Miller. 1993 (in press). "Optimizing Bioventing in Shallow Vadose Zones in Cold Climates." *Hydrological Sciences Journal 38*(4).

Looney, B. B., T. C. Hazen, D. S. Kaback, and C. A. Eddy. 1991. *Full Scale Field Test of the In-Situ Air Stripping Process at the Savannah River Integrated Demonstration Test Site*. WSRC-91-22. Westinghouse Savannah River Company, Aiken, SC.

MacDonald, T. R., and P. K. Kitanidis. 1993. "Modeling the Free Surface Due to a Recirculation Well." Poster presentation at the Second International Symposium on In Situ and On-Site Bioreclamation, San Diego, CA, April 5-8.

Miller, R. N., C. C. Vogel, and R. E. Hinchee. 1991. "A Field-Scale Investigation of Petroleum Hydrocarbon Biodegradation in the Vadose Zone Enhanced by Soil Venting at Tyndall AFB, Florida." In R. E. Hinchee and R. F. Olfenbuttel (Eds.), *In Situ Bioreclamation*. Butterworth-Heinemann, Stoneham, MA. pp. 283-302.

Raymond, R. L.   1976.   "Beneficial Stimulation of Bacterial Activity in Groundwater Containing Petroleum Hydrocarbons." *AICHE Symposium Series 73.* pp. 390-404.

Raymond, R. L.   1974.   "Reclamation of Hydrocarbon Contaminated Waters." U.S. Patent Office 3,846,290.

ENHANCING BIODEGRADATION WITH IN SITU AIR SPARGING: A CONCEPTUAL MODEL

R. L. Johnson

ABSTRACT

In most saturated subsurface environments, aerobic degradation is limited by the availability of oxygen. As a result, a number of techniques have been developed to increase oxygen supply in groundwater. One of the earliest proposed approaches was the direct injection of air into the porous medium (in situ air sparging). Simple calculations show that, if oxygen in the form of air is added to the groundwater upgradient of a contaminated zone, the limited carrying capacity of the water will not result in a significant level of degradation. On the other hand, if the oxygen can be delivered to zones where active biodegradation is occurring, large quantities of oxygen may be available for biodegradation. The central issue then becomes how effectively the oxygen can be transferred from the air to the water during the sparging process. At present no good data are available on oxygen transfer during sparging. In fact, there is a great deal of confusion concerning how air moves through saturated porous media. However, a conceptual examination of the air-sparging process combined with simple calculations suggest that the transfer of oxygen to the groundwater may be the limiting step in the sparging process.

INTRODUCTION

In the last few years the in situ air sparging (IAS) technique has become popular as a means of directly stripping contaminants out of the groundwater (Felton et al. 1992, Johnson et al. 1992, Leonard & Brown 1992, Marley et al. 1992). The direct injection of air into contaminated groundwater was proposed as early as 1974 as a means of increasing biodegradation of dissolved hydrocarbon contaminants (Raymond 1974).

One primary limitation of the IAS approach to enhancing biodegradation is that oxygen is only sparingly soluble in groundwater. As discussed by Yaniga and Smith (1984) and others, the limitation posed by a solubility of 10 mg/L means that only a small mass of oxygen can be present in the groundwater at any point in time. Yaniga and Smith also note that fouling/plugging of the sparging points is a limitation to the IAS technique. The solubility limitation can be overcome, at least in theory, if IAS occurs within the zone where biodegradation is active. In this case, the biodegradation process can be a continual sink for the oxygen and the process of oxygen transfer can continue. This is particularly desirable, because the volume of oxygen that can be introduced by the IAS process potentially is very large. Therefore, it is likely that the effectiveness of IAS for enhancing biodegradation will be controlled by the interaction of biodegradation and oxygen transport to groundwater.

To discuss the processes that are important to the success of IAS, it is useful to construct a hypothetical example. The example of a gasoline spill into a sandy porous medium is used here. When a gasoline spill occurs, the gasoline moves down through the unsaturated soil as a nonaqueous-phase liquid (NAPL). When it reaches the vicinity of the water table, its downward movement is slowed and it begins to spread laterally. Physical model examples of this process can be seen in the work of Schwille (1984) and others. A schematic representation of the distribution of gasoline following such a release is seen in Figure 1a.

In most field cases there will be some vertical movement of the water table as the result of either seasonal fluctuations or pumping. This causes some "smearing" of the gasoline in the medium (Figure 1b). This mechanism is important in ultimately distributing the gasoline as an "immobile residual" in the saturated medium. Once the bulk of the NAPL movement has stopped, the primary means of mass transport away from a residual source in the saturated zone will be by dissolution. For this example, it will be assumed that the gasoline spill occurred over an extended period and that the water table fluctuates 2 m per year (i.e., ±1 m), so on the average 1 m of product is trapped below the water table. For simplicity, it will be assumed that 1 m of the residual product is always below the water table. The soil that is contaminated below the water table may contain 50 L/m^3 of gasoline. If the spill produces a pool of product at the water table which is 10 m × 10 m = 100 m^2, then a 1-m thick zone of residual would contain 50 L/m^3 × 100 m^3 = 5,000 L of gasoline, or about 4,000 kg of gasoline trapped as residual within the saturated zone.

Dissolution Process

If the composite solubility of gasoline is about 100 mg/L or 100 g/m³, the groundwater flow is 0.3 m/d, and the porosity is 0.33, then the volume of water flowing through a 1 m x 10 m cross section perpendicular to the groundwater flow direction is ~1 m³/d or 365 m³/year. That means the theoretical amount of gasoline that could dissolve would be ~40 kg/year. At that rate, 100 years would be required to dissolve the gasoline. This analysis assumes that the vertical dispersion process is insignificant. There is now a considerable body of evidence showing that vertical dispersion is on the order of molecular diffusion, thus this is likely to be the case (McCarthy & Johnson 1992, Rivett & Cherry 1991).

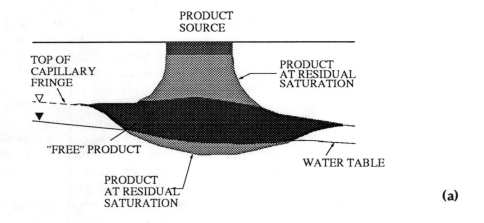

(a)

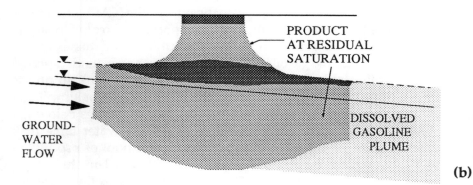

(b)

FIGURE 1. (a) Schematic drawing of a gasoline release in sand. (b) Schematic drawing of a gasoline release after a water table rise.

Given the very long time frames required for dissolution, the implication of the preceding example is that, although it is important to contain aqueous groundwater plumes, it is essential to address the source of the groundwater contamination (i.e., the NAPL) if the problem is to be eliminated within a reasonable time frame. As discussed below, this means that the remediation activities (e.g., sparging, degradation, and dissolution) must occur within the zone where the NAPL is present.

Biodegradation Process

The volume of oxygen delivered to the residual zone by groundwater flow for the example used above can be calculated in a manner similar to that for the dissolution calculation. As calculated above, 365 m^3 of water per year flows through the residual. If it is assumed that the upper limit for oxygen solubility is 10 mg/L or 0.01 kg/m^3, then ~4 kg of dissolved oxygen will flow through the residual zone each year (again assuming no vertical dispersion). If a degradation stoichiometry of 3 g O_2 to degrade 1 g of hydrocarbon is assumed, the oxygen delivered by the groundwater can degrade only 1.33 kg of hydrocarbon in a year. Thus it is likely that hydrocarbon-free, oxygenated water entering the upgradient end of the residual zone will quickly become depleted of oxygen and saturated with hydrocarbons, and that aerobic degradation will cease.

Sparging Process

As mentioned above, if the goal of the remediation process is to enhance biodegradation in order to remove the source of groundwater contamination (i.e., the zone of residual gasoline below the water table), it will be necessary to stimulate biodegradation within the residual zone. For this to be effective, sufficient oxygen must be delivered within that zone to allow biodegradation to proceed. It is clear that sparging upgradient of the residual zone will not be very effective, because even if oxygen concentrations increase to saturation, that does not result in a significant mass of oxygen into the groundwater. Similarly, if sparging occurs downgradient of the residual, the groundwater can be replenished with oxygen and degradation can occur within the plume, but it will not have any impact on dissolution within the source. (Some of the dissolved contaminants also may be removed by sparging.) If the contaminated groundwater has a significant residence time within the sparging zone, a good portion of the mass in the aqueous plume will be stripped out or degraded. This is, of course, desirable, but it does not increase the rate of removal of the source.

If sparging takes place within the residual zone, two things can happen simultaneously: (1) stripping of volatile contaminants from the groundwater will occur at the same time as dissolution, thus the amount of mass that can be dissolved per unit time will increase; and (2) oxygen added to the groundwater will allow biodegradation to proceed at the same time as dissolution, thus the amount of mass which can be dissolved per unit time will increase. Biodegradation may also be enhanced because sparging may reduce aqueous concentrations of contaminants to below inhibitory levels. Clearly, IAS within the residual zone is potentially a desirable approach. This gives rise to the question of how effective the IAS process can be at delivering oxygen to the groundwater zone.

As described above, during in situ sparging the air is injected directly into the medium and travels up through the medium to the unsaturated zone. The process of air movement through saturated porous media is not well understood at this time. It has been discussed briefly by Johnson et al. (1993). Basically, air moving through most porous media will travel in continuous air channels rather than as bubbles (Figure 2a). The channels themselves are likely to be small (<< 0.01 m) and widely separated (e.g., 0.1 to 1 m between them). In addition, mass transfer of oxygen to the groundwater likely will be controlled by diffusion between the channels and the water. Similarly, hydrocarbon transport from groundwater to the channels likely will be controlled by diffusion. It is also important to recognize that minor heterogeneities can dramatically effect the distribution of air in saturated porous media. An example of this is shown schematically in Figure 2b.

It is possible to make some rough estimates of the maximum degradation rates that are possible in groundwater for the IAS process. To accomplish this it will be assumed that the air channels have a radius of 0.2 cm, that they are on the average 20 cm apart, and that their length through the residual zone is 100 cm. As shown in Figure 3, it also will be assumed that the oxygen concentration decreases from 10 mg/L to 0 over a distance of 1 cm from the channel. Using the above values it is possible to calculate the daily flux of oxygen into the medium from channels within a 100 m^2 residual zone using the equation (Crank 1956):

$$Q_t = \frac{2\pi Dt\,(C_2 - C_1)}{\ln\,(b/a)}$$

where D = diffusion coefficient = 0.00001 cm^2/s
       t = time = 86,400 sec
       C_2 = concentration at the channel/water interface = 10 mg/L
       C_1 = 0

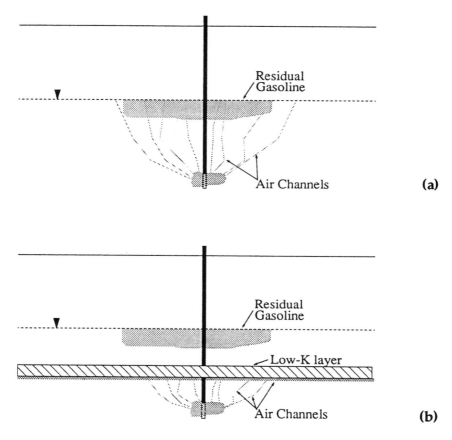

FIGURE 2. (a) Schematic drawing of airflow during in situ air sparging in homogeneous sand. (b) Schematic drawing of airflow during in situ air sparging in heterogeneous sand.

b = thickness of the oxygen gradient zone plus the channel = 1.2 cm

a = radius of the channel = 0.2 cm

If equation 1 is multiplied by the length of the channel and the number of channels within the residual (100 cm and 2,500, respectively), the total flux of oxygen into the groundwater can be calculated. For the above example the flux is approximately 7.5 g/day. This translates to a degradation rate of approximately 2.5 g of hydrocarbon per day, which is insignificant. Of course, there are many assumptions inherent in this calculation, including the size and spacing of the channels and the thickness of the

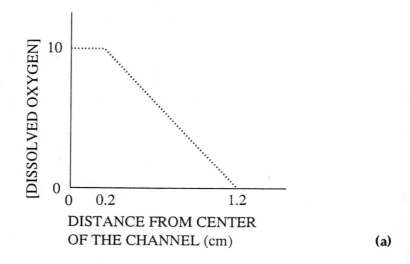

(a)

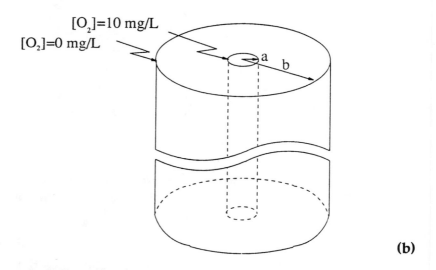

(b)

FIGURE 3. (a) Schematic drawing of oxygen concentration as a function of distance from the center of a channel. (b) schematic drawing of a channel surrounded by a zone containing oxygen.

oxygen gradient zone. If mixing at the interface between the channels and the water is greater than predicted by diffusion, the mass transfer of oxygen to the groundwater may be significantly greater. Unfortunately, there are currently no experimental data that have examined this process.

If the transfer of oxygen to the groundwater is not limiting, there is undoubtedly some limit beyond which the degradation rate cannot be increased. If it is assumed that within the residual zone we can maintain 10 mg/kg/day degradation rates, and the residual zone has a volume of 100 m³, then ~1.8 kg/day or 660 kg/year could be degraded. Coupled with the amount that can be stripped by the sparged air, this translates to a significant mass removed. The dramatic difference in degradation rate between the last two examples points to the need to better understand the oxygen transfer process during IAS.

SUMMARY

The IAS process has the potential to deliver a large mass of oxygen to the groundwater. The key to the success of IAS as a technique for remediating NAPL contamination may be in the ability to deliver the oxygen to where it is most useful, specifically to the immediate vicinity of the NAPL. Given the current understanding of how NAPLs distribute themselves within the subsurface, it should be possible to identify the zones to be sparged. The next task is to deliver air to the NAPL-containing zone(s). Given the lack of understanding with regard to air movement in saturated media, this is an area in which additional research needs to be done. Finally, there is the uncertainty associated with the efficiency with which oxygen is transferred to the groundwater. The examples discussed here suggest that these issues must be addressed if the IAS technique is to be proven effective.

REFERENCES

Crank, J. 1956. *The Mathematics of Diffusion*, p. 414, Oxford University Press, London.

Felton, D. W., M. C. Leahy, L. J. Bealer, and B. A. Kline. 1992. "Case Study: Site Remediation using Air Sparging and Soil Vapor Extraction." In *Proceedings of the 1992 Petroleum Hydrocarbons and Organic Chemicals in Ground Water: Prevention, Detection and Restoration*, pp. 395-412, November 4-6, 1992, Houston, TX.

Johnson, R. L., W. Bagby, M. Perrott, and C. Chen. 1992. "Experimental Examination of Integrated Soil Vapor Extraction Techniques." In *Proceedings of the 1992 Petroleum Hydrocarbons and Organic Chemicals in Ground Water: Prevention, Detection and Restoration*, pp. 441-452, November 4-6, 1992, Houston, TX.

Johnson, R. L., P. C. Johnson, R. Hinchee, D. B. McWhorter, and I. Goodman. 1993. "A Review of Field Data for In-Situ Air Sparging." Submitted to *Ground Water Monitoring and Remediation.*

Leonard, W. C., and R. A. Brown. 1992. "Air Sparging: An Optimal Solution." In *Proceedings of the 1992 Petroleum Hydrocarbons and Organic Chemicals in Ground Water: Prevention, Detection and Restoration,* pp. 349-364, November 4-6, 1992, Houston, TX.

Marley, M. C., D. J. Hazelbrouck, and M. T. Walsh. 1992. "The Application of In-Situ Air Sparging as an Innovative Soils and Groundwater Remediation Technology." *Ground Water Monitoring Review* 12: 135-145.

McCarthy, K. A., and R. L. Johnson. 1992. "The Transport of Volatile Organic Compounds Across the Capillary Fringe." In press *Water Resources Research.*

Raymond, R. L. 1974. "Reclamation of Hydrocarbon Contaminated Waters." U.S. Patent Office 3,846,290.

Rivett, M. O., and Cherry, J. A. 1991. "The Effectiveness of Soil Gas Surveys in the Delineation of Groundwater Contamination: Controlled Experiments at the Borden Field Site." In *Proceedings of the Conference on Petroleum Hydrocarbons and Organic Chemicals in Groundwater: Prevention, Detection and Restoration,* pp. 107-124, National Water Well Association, Houston, Texas, November 20-22.

Schwille, F. 1984. *Dense Chlorinated Solvents in Porous and Fractured Media.* Translated by J. F. Pankow. Lewis Publishers, Chelsea, MI. p. 146.

Yaniga, P. M., and W. Smith. 1984. "Aquifer Restoration Via Accelerated Insitu Biodegradation of Organic Contaminants." In *Proceedings of the 1984 Petroleum Hydrocarbons and Organic Chemicals in Ground Water: Prevention, Detection and Restoration,* pp. 451-470, November 5-7, 1984, Houston, TX.

AIR SPARGING: AN EFFICIENT GROUNDWATER AND SOILS REMEDIATION TECHNOLOGY

M. C. Marley and F. Li

ABSTRACT ⎯⎯⎯⎯⎯⎯⎯⎯⎯⎯⎯⎯⎯⎯⎯⎯⎯⎯⎯⎯⎯

Air sparging is an innovative methodology for remediating organic compounds present in contaminated, saturated soil zones. In the application of the technology, sparging wells are used to inject a hydrocarbon-free gaseous medium into the saturated zone below or within the areas of contamination. Volatile organic compounds are dissolved in the groundwater and sorbed on the soil partition into the advective air phase, effectively simulating an in situ air stripping system. The stripped contaminants are transported in the air phase to the vadose zone, generally within the radius of influence of a standard vapor extraction. Under optimal environmental conditions, enhanced biodegradation of volatile and semivolatile organic compounds may be attained through the oxygenation of the groundwater. Air sparging is a complex, multi-fluid phase process. Major design considerations include site geology, contaminant type, gas injection pressures and flow-rates, injection interval (areal and vertical), and site-specific parameters that dictate the feasibility of enhanced biodegradation. In this study, a mathematical model was developed to simulate the airflow field during the sparging process and to examine the limitations imposed by the site geology. The development of the model is described and model simulations are performed. Model simulations and field-collected data are compared for two sites.

INTRODUCTION

Accidental releases of organic compounds (OCs) into the subsurface environment in the form of petroleum products or industrial solvents can result in costly remediation. Although virtually any form of remediation

1-56670-084-1/94/$0.00 + $.50

is expensive, developing a well-planned, cost-effective strategy at the onset of a spill or release can minimize expenses that accumulate throughout the duration of a cleanup project. Removal of the OC source is the primary consideration to ensure effective remediation. Soil contamination underlying and in the vicinity of a leaking underground storage tank (LUST) or a surface spill is a potential long-term contributor/source to the migration of hazardous vapors in vadose zone soil and to dissolved OCs in groundwater. Frequently, contaminated soils exist below the groundwater table (GWT) when free-phase product mounds on the GWT and is transported vertically in response to seasonal GWT fluctuations or drawdown from pumping in nearby groundwater/product recovery wells. Dense, nonaqueous-phase liquids (DNAPLs) frequently are found on soils below the GWT as globules and/or residuals, due to their density-driven vertical transport.

A few commercially applicable in situ remediation technologies exist that can be applied as remedial alternatives at OC spill sites, although generally no one technique can accomplish all the objectives of a complete site cleanup. Using pump-and-treat methods to remediate OCs sorbed and/or trapped in saturated zone soils is considered to have significant limitations (Mackay & Cherry 1989) due mainly to standard pump-and-treat system designs, site-specific soil heterogeneities, contaminant distribution, and kinetic limitations to the mass removal process. Techniques such as soil washing and augmented biodegradation provide potential enhancements to pump-and-treat-based remediation methods.

Soil vapor extraction (SVE) and bioventing have been demonstrated to be successful and cost-effective remediation technologies for removing OCs from vadose zone soils. These techniques involve the controlled application of an air pressure gradient to induce an airflow through soils contaminated with OCs. As soil gas is drawn toward the vacuum source (vapor extraction well), the equilibrium both between the OC phases as product, and in the soil between the soil vapor and soil moisture, is upset, causing enhanced partitioning into the vapor phase. OCs in the vapor phase subsequently are removed from the subsurface and treated with a standard vapor extraction off-gas treatment system(s). The oxygenation of the vadose zone also promotes the aerobic biodegradation of volatile and semivolatile OCs. One limitation of using SVE and/or bioventing is that neither is an optimal remediation technology for addressing contaminated soils existing below the GWT.

A number of techniques have been developed and employed to expand the SVE process to include effective remediation of OCs in saturated zone soils. Artificial water table drawdown is one approach that may be used to expose contaminated soils in the saturated zone to the advective air phase, thereby increasing the efficiency of the SVE process. However, in some cases, artificial water table drawdown is neither practical nor cost effective. An innovative, alternative approach is the application of air sparging technology, also referred to as soil/groundwater aeration, to inject a hydrocarbon-free gaseous medium (most commonly air) into the saturated zone below/within the areas of contamination. With air sparging, the OC contaminants dissolved in the groundwater and sorbed/trapped on the soil may partition into the advective gaseous phase to effectively simulate an in situ, saturated zone air stripping system. The stripped contaminants subsequently are transported in the air phase to the vadose zone, within the radius of influence of an operating soil vapor extraction system. The contaminant vapors are drawn through the vadose zone to the vapor extraction well(s) and are treated using standard vapor extraction off-gas system(s). A schematic depicting a typical air sparging system is presented in Figure 1.

Limited references exist in the literature as to the design and/or success of the laboratory or field application of air sparging. Apparently the process was first used as a remediation technology in Germany in the mid-1980s, predominantly to enhance the cleanup of chlorinated solvent-contaminated groundwater (Gudemann & Hiller 1988). More

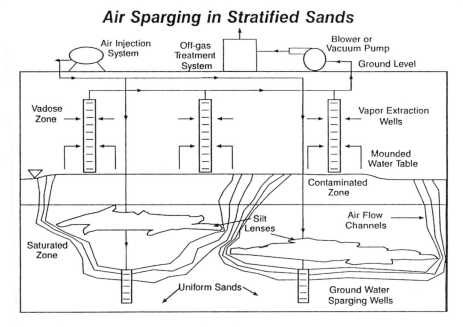

FIGURE 1. Typical sparging system configuration.

recently the technology has been used in the enhanced remediation of gasoline-contaminated saturated zone soils and groundwater (Ardito & Billings 1990, Brown et al. 1991, Marley 1991). Apparently, in each of these cases the design of the air sparging systems has been empirically based.

The authors of this article have performed air sparging field pilot tests and implemented full-scale SVE/air sparging systems on numerous sites across the United States. Experience developed on these projects has demonstrated that numerous important criteria must be considered when designing, installing, and operating an air sparging system, not only to ensure effective remediation of saturated zone soils/groundwater, but also to preclude or control potentially displaced and mobilized hazardous soil gas vapors, free-phase product, or dissolved-phase contaminants in the saturated zone. To better understand the technology, mathematical models describing the process have been developed.

DESCRIPTION OF THE MATHEMATICAL MODEL

Mathematically, one of the difficulties in modeling the air sparging process is that the governing equations (fluid flow in variably saturated porous media) traditionally are coupled, highly nonlinear partial differential equations. In addition, the highly nonlinear fluid saturation/capillary pressure and capillary pressure/relative permeability relationships present a difficult computational problem, even when using numerical analysis techniques.

Some basic assumptions and simplifications were used to set up a practical design model. First, the influence of air injection from line or point sources was observed and was assumed to have very little effect on the regional groundwater flow field. The water table level is assumed to remain at or close to its original level under steady-state airflow conditions. The air sparging problem is then considered as a flow of air and contaminants, and water flow is neglected. In contrast, the water flow induced by local pumping wells can affect the airflow significantly. The steady-state governing equations for the air and water flow fields are as follows:

For water flow equations:

$$\nabla \bullet K_{rw} \; K_{ws} \; \nabla \bullet P_w = 0 \qquad (1)$$

For airflow equations:

$$S + \nabla \bullet K_{ra} \; K_{as} \nabla \bullet P_a = 0 \tag{2}$$

where K_{rw} = Relative permeability for water
K_{ra} = Relative permeability for air
K_{ws} = Saturated water permeability
K_{as} = Saturated air permeability
P_w = Water pressure or potential
P_a = Air pressure
S = Air source term

For the well:

$$S = \frac{Q}{L} \; \delta \, (r - r^1) \tag{3}$$

where δ = Delta function
$\dfrac{Q}{L}$ = Airflow rate per unit length of screen

Increasing the degree of air saturation within a pore during the sparging operation causes an increase in the permeability of the pore to air. Researchers have performed many experiments in porous media and set up numerous models to find the relationships between relative permeability and capillary pressure and degree of saturation. Van Genuchten (1980; Van Genuchten et al. 1985, 1988, 1991) presented a closed form of Mualem's equation that defines permeability as a function of matrix potential and produced the following equations to describe permeability as a function of the degree of pore saturation. According to Van Genuchten (1980) and Van Genuchten et al. (1991):

$$P_c = P_w - P_a = (\rho g / \alpha) \, ((S_e^{1/m}) - 1) \tag{4}$$

$$S_e = (S_w - S_{rw}) / (1 - S_{rw}) \tag{5}$$

$$S_w + S_a = 1 \tag{6}$$

$$K_{rw} = S_e^{\frac{1}{2}} (1 - (1 - S_e^{1/m})^m)^2 \tag{7}$$

where P_c = Capillary pressure
α, n = Soil constants (determined empirically or through laboratory testing)
S_e = Effective saturation

m = 1 − 1/n
S_w = Degree of water saturation
S_{rw} = Irreducible water saturation
S_a = Degree of air saturation

Also, according to Brooks and Corey (1966):

$$K_{ra} = (1 - S_e^2)(1 - S_e^{(2+\lambda)/\lambda})   \tag{8}$$

where λ = Empirical parameter

The movement of air from an injection well under the action of its implied pressure in the saturated zone is similar to the flow of water in the unsaturated zone. For the latter case, many researchers have reasonably modeled this complex process. It is possible to use ideas from these models to help simulate the air sparging process. First, when air is injected into a water-saturated porous medium, air channels are formed and air will tend to move up under buoyancy and applied pressure forces. For a confined aquifer, where the water pressure is approximately equal everywhere, the size of the air bubble or channel should not change during its upward motion. The air will tend to move horizontally on the top of the aquifer. For an unconfined aquifer, with the decrease in water pressure at a higher elevation, the size of the air channel has a tendency to increase. Therefore, inside a flow domain, various airflow patterns may exist simultaneously. When air traverses horizontally and vertically through the soil column, an air-filled space is created within which contaminant volatilization can occur. During this process, the contaminants transfer into the gas phase and are moved within the air to the vadose zone. Mass transfer occurs across the interphase boundaries (Marley 1991, Marley et al. 1992). At the same time, oxygen is transferred from the injected air with a resultant oxygenation of the groundwater.

Based on the above concepts, the following assumptions were made in developing the numerical model:

1. At steady-state flow, the rise in the groundwater table is neglected.
2. Sparging wells are represented as line sources.
3. Airflow is assumed to obey Darcy's law.

To carry out the computations, a boundary element groundwater flow software, FLWROC6 (Medina & Ligget 1988), was modified. The air sparging model was developed, and the model the nonlinear characteristics were determined through an iteration process from an initial estimate of air permeability.

MODEL SIMULATIONS

Model Verification

To verify the computational accuracy of the three-dimensional model, the results of the model simulations were compared with those of a three-dimensional analytical solution for a simple homogeneous domain. The assumed computation domain is a long rectangular box. The dimension of the domain is $0 < X = Y < 200$, $0 < Z < 10$. The boundary condition at $X = 0$ and 200 was set at a constant air pressure. All the other four faces have Neumann conditions. The injection well is located at the center of the domain. Air permeability and total porosity were estimated at 2×10^{-8} cm^2 and 0.35, respectively. The parameters taken from the literature for Van Genuchten's equations are: $x = 0.145$, $S_{rw} = 0.40512$, and $n = 2.7$. Figure 2 shows the comparison between the numerical and analytical model results for simulations at various flowrates.

Model Simulation (Theoretical)

Simulations were performed on four different hypothetical soil profiles. The dimensions of the domain were $300 \times 100 \times 20$ (unit-less). On the

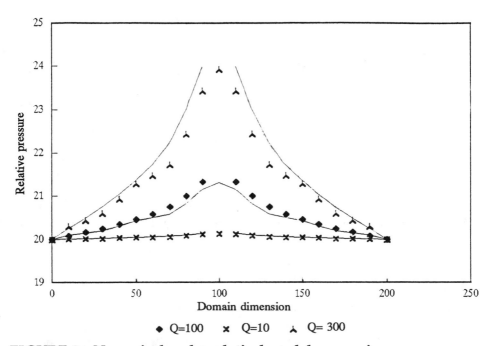

FIGURE 2. Numerical and analytical model comparison.

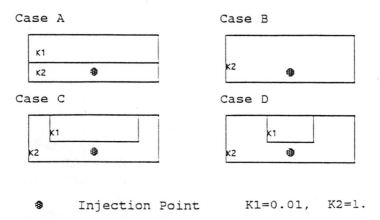

⊛   Injection Point   K1=0.01,   K2=1.

FIGURE 3. Model simulation test cases.

top face the air pressure was fixed, and for all the other faces the air flux was set to zero. Figure 3 presents the four soil profiles modeled. Figure 4 shows the distribution of air pressure along an observation line from point $X = 0$, $Y = 50$, $Z = 7$ to point $X = 300$, $Y = 50$, $Z = 7$. Intuitively, the relative magnitude of airflow under the air sparging process for each soil profile appears appropriate. A degree of computational error exists based on the grid spacing and the assumption of a line source.

Model Simulation (Field Studies)

The sparging model was applied to simulate air sparging at two job sites.

Site 1: Wisconsin Project. This site in Wisconsin is located at an active feed and grain distribution center. Both saturated and vadose zone soils have been contaminated with gasoline-range OCs from several leaky underground storage tanks (LUSTs).

Soils in the vicinity of the site have been identified as a silt loam. The silt loam is characterized as a poorly drained soil of the type that generally forms in the area of old lake beds. It typically consists of black loam grading into loamy, very fine sands. The soil is moderately permeable and has a seasonal high water table that often comes to within 1 ft (0.3 m) of the ground surface.

The contamination covers an area of approximately 60 × 80 ft (18 × 24 m). Soils in the vicinity of the air injection wells were removed during tank excavations and replaced with fill material. The soil strata, in general terms, consist of fill material down to 2.0 ft (0.6 m) below grade. From

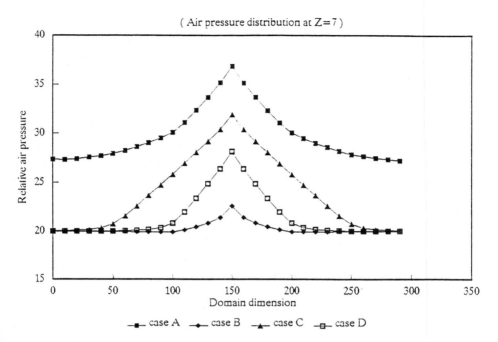

FIGURE 4. Results of numerical model simulation.

2.0 to 5.0 ft (0.6 to 1.5 m) there exists a layer of lean clay and below that are several mixed layers of sandy silt and clay from 5.0 to 10.0 ft (1.5 to 3.0 m). Finally, soils present from 10.0 to 19.0 ft (3.0 to 5.8 m) below grade consist of fine sands.

Preliminary subsurface airflow modeling estimated air permeabilities for these soils to be in the range of approximately 1×10^{-8} to 1×10^{-7} cm^2.

The average depth to groundwater across the air sparging area is 8 ft (2.4 m) below grade. The groundwater gradient is shallow across the site. Drawdown field tests were performed and indicated that the water-bearing stratum is very permeable. Hydraulic conductivities determined from tests in the immediate vicinity of the sparging well ranged from 1×10^{-1} cm/sec to 5×10^{-3} cm/sec.

The dimensions of the mathematical domain were set at 100 × 70 × 10 ft (30 × 21 × 3 m), and the entire domain surface was discretized into 98 triangular elements in the computation. The sparging well was located at x = 25 ft, y = 35 ft (x = 7.6 m, y = 10.7 m), and the well was screened at 4 ft (1.2 m). The airflow rate was 3.2 cfm. On the top of the domain, the air pressure was fixed at atmospheric pressure, and on the other faces, the airflow rate was set at zero. Initially, air saturation below the groundwater table was set close to zero (S_a = 0.0001). Based on the field boring log data, the initial estimate of air permeability was set at 2×10^{-8} cm^2; 30 observation points were set at equal intervals along line A (line A begins at point x = 0, y = 35, z = 4.9 and extends to point x = 100, y = 35, z = 4.9). The distance between observation points was approximately 3.33 ft (1 m).

Figure 5 compares the model-generated and field-measured air pressure in the subsurface soils at the site. Figure 6 shows a perspective view of the air pressure distribution during the sparging event. Reasonable agreement was observed, and the computational error at the wellhead associated with the line source assumption is evident. Figure 7 presents the horizontal and vertical airflow velocities in the domain at Z = 4.9 ft (1.5 m) for an airflow rate of 3.2 cfm. Figure 8 presents an evaluation of the effects of applying different well screen intervals on the vertical velocity profile. From these results, the radius of influence of the sparging well and the average air permeability in the soil brought about by this flowrate were calculated as 18 ft (5.5 m) and 0.276×10^{-9} cm^2, respectively.

Site 2: New Jersey Project. This site in New Jersey is a former industrial manufacturing facility. Both saturated and vadose zone soils have been contaminated with OCs, primarily toluene. Small amounts of chlorinated solvents also have been detected at the site.

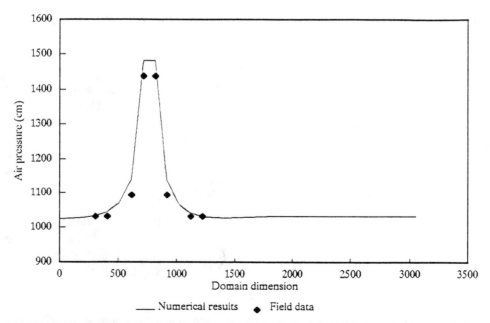

FIGURE 5. Comparison of numerical model and field data (Wisconsin).

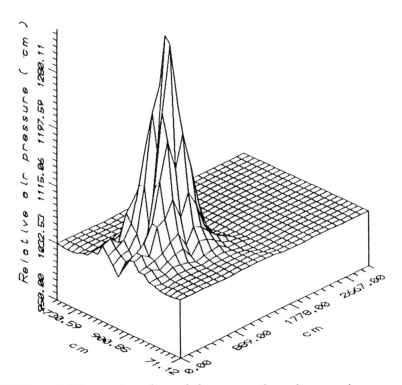

FIGURE 6. 3-Dimensional model perspective air sparging.

The approximate area of groundwater contamination is 67,000 ft² (6,225 m²). Soils across the site consist of medium to coarse sands to a depth of at least 20 ft (6 m) below grade. The hydraulic conductivity ranges between 0.0114 and 0.214 ft/min (0.21 and 3.91 m/s). The air permeability of the vadose zone is approximately 1×10^{-6} cm².

The computational domain was divided into two areas, and the sparging program was run for Area 2. The dimensions were set as shown on Figure 9. This is an irregular domain with the following boundary conditions: P_a equals atmospheric pressure on the top face, and on the other faces, the air flux is set to zero.

The initial estimate for the air permeability was 1×10^{-8} cm². After five iterations, the computational error was smaller than 3%. The results along observation line B (x = 800 cm, y = 500 cm, and x = 2350 cm, y = 1400 cm, at z = 8 cm) are shown in Figure 10. The distance between observations points is 0.72 ft (0.22 m).

Based on these results, the average air permeability was 0.26×10^{-9} cm² under the test airflow rate, and the effected radius of influence of the air sparging well was about 25 ft (7.6 m).

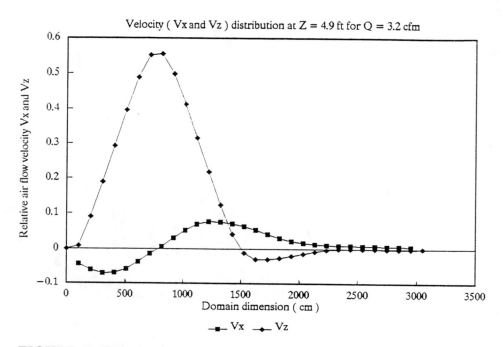

FIGURE 7. Results of numerical model simulation.

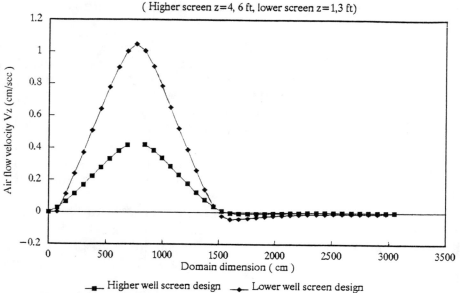

FIGURE 8. Comparison of two design modeling results.

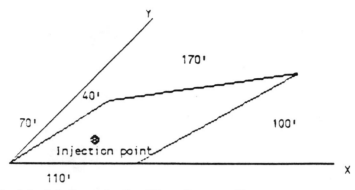

FIGURE 9. Model domain for New Jersey site.

CONCLUSIONS

The importance of developing a better understanding of the mechanisms controlling the air sparging process is evident. To aid in this understanding, preliminary mathematical models have been developed to simulate the process.

In this study, a modified three-dimensional boundary element model was developed to simulate the distribution of the air pressure and the air velocity vectors during sparging (i.e., the flow field is simulated).

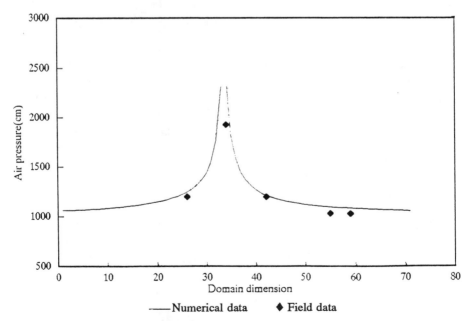

FIGURE 10. Comparison of numerical model and field results (New Jersey).

The numerical model agrees reasonably well with the results of both analytical modeling of the air sparging process and with limited field data collected from pilot testing of the technology. Additional comparisons are needed to refine the capabilities of the model. Further model simulations will provide insight into the sensitivity of the parameters required as input to the model.

There are a few disadvantages to this particular model. The first is that the numerical data generated near the line source (i.e., near the well screen) are not correct since the potential is assumed infinite at the source. This error is due to the mathematical assumption. According to the limited modeling done to date inside the range of 3% of the domain dimension at the well, the data are not as reliable. A second disadvantage is that a powerful PC or a workstation is needed to run significant numbers of elements and iterations.

REFERENCES

Ardito, C. P., and J. F. Billings. 1990. "Alternative Remediation Strategies: The Subsurface Volatilization and Ventilization System." *Proceedings of The Conference on Petroleum Hydrocarbons and Organic Chemicals in Ground Water: Prevention, Detection and Restoration*, NWWA, pp. 281-296.

Brooks, P. M., and A. T. Corey. 1966. *Hydraulic Properties of Porous Media*, Hydrol. Paper 3. University of Colorado, Fort Collins, CO.

Brown, R., C. Herman, and E. Henry. 1991. "The Use of Aeration in Environmental Cleanups." *Proceedings of Haztech International Pittsburgh Waste Conference*.

Gudemann, H., and D. Hiller. 1988. "In Situ Remediation of VOC Contaminated Soil and Groundwater by Vapor Extraction and Groundwater Aeration." *Proceedings of the Third Annual Haztech International Conference*, Cleveland, OH.

Luckner, L., M. Th. Van Genuchten, and D. R., Nielsen. 1989. "A Consistent Set of Parametric Models for the Two-Phase Flow of Immiscible Fluids in the Subsurface." *Water Resource Research*. 25(10): 187-219.

Mackay, D. M., and J. A., Cherry. 1989. "Ground Water Contamination: Limitations of Pump and Treat Remediation." *Environmental Science and Technology*, 23(6): 630.

Marley, M. C. 1991. "Air Sparging in Conjunction with Vapor Extraction for Source Removal at VOC Spill Sites." *Proceedings of the Fifth National Outdoor Action Conference on Aquifer Restoration, Groundwater Monitoring and Geophysical Methods*, pp. 89-103.

Marley, M. C., D. J., Hazebrouck, and M. T. Walsh. 1992. "The Application of In-Situ Air Sparging as an Innovative Soils and Groundwater Remediation Technology." *Groundwater Monitoring Review* 12(2): 137-145.

Medina, D. E., and J. A. Ligget. 1988. "Three-Dimensional Boundary-Element Computation for Potential Flow in Fractured Rock." *International Journal for Numerical Methods in Engineering* 26: 2319-2330.

Parker, J. C., R. J. Lenhard, and T. Kuppuscanny. 1987. "A Parametric Model for Constitutive Properties Governing Multiphase Flow in Porous Media." *Water Resource Research*, Vol. 23, 4th ed., pp. 618-624.

Sleep B. E., and J. F., Sykes. 1989. "Modelling the Transport of Volatile Organics in Variably Saturated Media." *Water Resource Research* 25(1): 81-92.

Van Genuchten, M. Th. 1980. "A Closed Form Equation for Predicting the Hydraulic Conductivity of Unsaturated Soils." *Soil Science Society, AM, J.,* 44: 892-898.

Van Genuchten, M. Th., F. Kaveh, W. B. Russel, and S. R. Yates. 1988. "Direct and Indirect Methods for Estimating the Hydraulic Properties of Unsaturated Soils." *Proceedings of a Symposium Organized by ISSS,* pp. 61-72.

Van Genuchten, M. Th, F. J. Leij, and S. R. Yates. 1991. "The RETC Code for Quantifying the Hydraulic Functions of Unsaturated Soils." June 1991.

Van Genuchten, M. Th., and D. R. Nielsen. 1985. "On Describing and Predicting the Hydraulic Properties of Unsaturated Soils." *Annales Geophysicae 3(5):* 615-628.

USE OF AIR SPARGING FOR IN SITU BIOREMEDIATION

R. A. Brown, R. J. Hicks, and P. M. Hicks

ABSTRACT

A new technology that is having an impact on in situ bioremediation is air sparging. Air sparging is the injection of air, under pressure, directly into the formation below the water table. This technique creates a transient air-filled porosity that enhances the oxygenation of the groundwater. Air sparging provides more oxygen at significantly lower cost than older technologies such as hydrogen peroxide. Air sparging appears to be simple technology, but is complex. For example, while serving to oxygenate water and to remove volatile compounds, air sparging can also displace water and vapors. If uncontrolled, this displacement can lead to the undesired migration of contaminants. As a result, care must be given to the design and installation of an air sparging system. This paper discusses the benefits of air sparging, its limitations, and principles of system design and operation. Because pilot testing is critical to the design of an air sparging system, this paper considers methods of pilot testing and the use of multiple parameters to monitor performance during the pilot test. The relative merits of parameters such as dissolved oxygen (DO) pressure, water table elevation, and changes in volatile organic compound (VOC) levels are discussed. Finally, this paper discusses the efficacy of air sparging as a primary oxygenation system for in situ bioremediation.

INTRODUCTION

Air sparging is a relatively new technology, having been practiced in the United States for 3 to 5 years. Air sparging is the direct injection of air, under pressure, into saturated matrixes. The benefit of air sparging is that it can engender the physical removal and/or biodegradation of many organic compounds. VOCs in soil or groundwater can be stripped/volatilized by the injected airstream. Degradable substances, volatile or not, are aerobically metabolized. This combined physical and biological removal process provides speed and flexibility.

Air sparging is fast becoming a preferred system of supplying oxygen for in situ bioremediation. Compared to earlier oxygenation systems, air sparging offers several advantages. It is generally more effective, less expensive, more benign, and easier to control than oxygenation methods such as in-well aeration, or the use of H_2O_2. Also, because of its low cost, and ease of application, air sparging has obviated the need for alternate electron acceptors such as nitrate.

Air sparging is, however, not without its limitations. A primary limitation is that, if the airstream is not properly controlled, air sparging runs the risk of physically spreading contamination through the potentially rapid and uncontrolled migration of nonaqueous-phase liquids (NAPLs), dissolved components, and/or vapors. Lack of control may be caused by geological factors such as layering, or by operational factors such as overpressurizing. Therefore, the keys to controlling the airstream and to its effective use are the proper selection of a site, and the proper design and installation of the system.

BACKGROUND

To understand the value of air sparging for bioremediation, one needs to understand the importance of oxygen supply to the development of bioremediation. Although bioremediation was first practiced on a commercial scale in 1972 to treat a gasoline pipeline spill in Ambler, Pennsylvania (Raymond 1976), it was never widely used until the mid-1980s. The first applications of commercial bioremediation attempted to stimulate indigenous bacteria by enriching the subsurface environment with oxygen, nitrogen, phosphorus and trace minerals. These early bioremediation systems used simple in-well aeration to supply oxygen. This method was extremely limited in its ability to supply oxygen. When early applications of bioremediation did not work effectively, the cause of failure generally was found to be a lack of sufficient oxygen (Floodgate 1973; ZoBell 1973).

Oxygen supply was, thus, identified as the central issue to be resolved if bioremediation technology was going to have general applicability. This focus on oxygen supply lead to the first major innovation in bioremediation — the use of H_2O_2 as an oxygen "carrier." H_2O_2 could potentially supply orders of magnitude more oxygen than in-well aeration because of its miscibility in water and its ability to supply ~0.5 ppm oxygen for each ppm added. Because of this potential for increasing oxygen availability, H_2O_2 had the promise of making bioremediation commercially viable. Its use was, however, limited by its cost, potential biotoxicity, and instability. H_2O_2 never fully satisfied the need for a generally applicable, effective, and inexpensive oxygen source.

At first, H_2O_2 was the sole oxygen source. It was injected into the vadose zone as a solution in nutrient-amended water to provide oxygen in both the vadose and saturated zones. H_2O_2 did provide a significant improvement in oxygen supply over simple in-well aeration, but its use for unsaturated soil treatment was quickly supplanted by soil vapor extraction (SVE) technology (Brown & Crosbie 1989). SVE, however, could not address contaminants below the water table and H_2O_2 remained the system of choice for saturated matrixes. Thus bioremediation systems became hybrid, integrated systems using both SVE and H_2O_2 (Brown et al. 1991).

The uncontrolled or premature decomposition of H_2O_2 and its toxicity to bacteria continued to be problems that limited the use of H_2O_2. These problems lead to the research of an alternative (non-oxygen) electron acceptor and of alternative means of supplying oxygen.

Nitrate has been the most common electron acceptor investigated as a substitute for H_2O_2 (Hutchins et al. 1991) because it is relatively inexpensive, is very soluble in water, is not adsorbed to soil matrixes, and does not decompose. However, nitrate's more facile distribution within an aquifer is offset by regulatory limitations and by its ineffectiveness with certain classes of compounds, such as aliphatic hydrocarbons or benzene.

Recently, air sparging has been used to provide oxygen below the water table (Brown & Jasiulewicz 1992) because it can distribute oxygen rapidly and somewhat uniformly across the entire site, and it is relatively inexpensive to implement and operate. Air sparging has the potential for providing the same benefits to saturated zone treatment that SVE provides in vadose zone treatment.

BENEFITS OF AIR SPARGING

The application of air sparging results in a complex series of removal processes: volatilization, biodegradation, and solubilization. In its application to bioremediation, the dominant and secondary removal mechanisms depend on the volatility of the contaminant. As shown in Figure 1, with a highly volatile contaminant, the primary removal mechanism is volatilization. Direct volatile removal of VOCs by air sparging can be

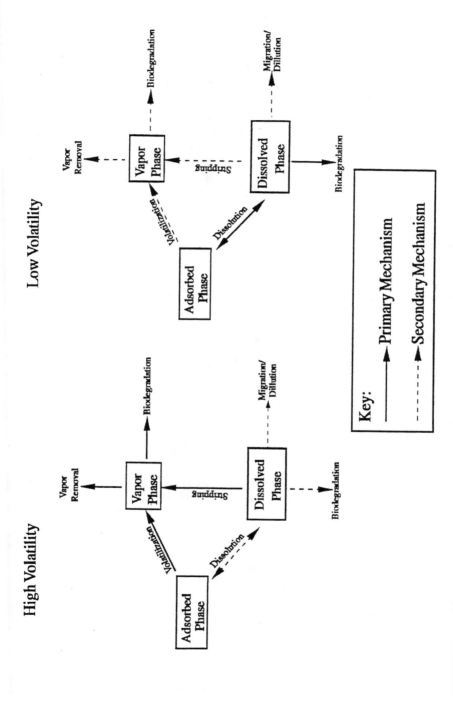

FIGURE 1. Air sparging partitioning and removal mechanisms as a function of volatility.

significant (Hiller & Gudemann 1988; Marley et al. 1992). By contrast, with a low-volatility contaminant, the primary removal mechanism is through biodegradation.

In bioremediation, air sparging effectively supplies oxygen to the saturated zone to enhance aerobic metabolism due to four factors. First, air is diffused throughout the entire sparged interval rather than only from the surface of the aquifer. Second the diffusive path length of oxygen in water is considerably shortened from ~100 to 1,000 mm (the depth of the contaminated interval below the water table) to 0.1 to 1.0 mm (the distance between air channels distributed in the aquifer), increasing the rate at which oxygen is supplied. Third, the "turbulence" caused by air movement enhances the dissolution and distribution of oxygen into the water phase by mixing the water column. Air sparging provides "totally mixed" reactor rather than a "plug-flow" reactor as with the injection of H_2O_2. Fourth, compared with in-well aeration, the use of H_2O_2, or the use of nitrate, air sparging can supply more oxygen equivalents for bioremediation, as shown in Table 1. Because biodegradation is critically dependent on oxygen supply, the efficient aeration and the increased oxygen availability engendered by air sparging enhance bioremediation.

Evidence for Enhanced Oxygenation

The evidence for enhancement of bioremediation by air sparging is both direct and indirect. The direct evidence is the consumption of O_2 or the increase in CO_2 observed during air sparging. Concentrations of CO_2 of 8 to 12% have been observed during air sparging of petroleum hydrocarbons at depths of 3.7 to 7.6 m below the water table (Leonard & Wilson 1992). The indirect evidence, which is often easier to obtain, is the increase in DO levels measured in groundwater during air sparging operations. Within 0.5 to 2 hours after sparging is initiated, DO levels in aquifers contaminated with petroleum hydrocarbons will often increase

TABLE 1. Oxygen availability.

	kg Carrier/ kg O_2 Equivalent	L Carrier/ 1,000 kg O_2 Equiv.
Aerated Water	100,000	1×10^8
H_2O_2 (1,000 mg/L)	2,200	2×10^6
Nitrate (10 g/L)	176	176,400
Air Sparging (20% O_2)	4.5	55,000

from 1 to 2 mg/L to saturation (>10 mg/L) (Leonard & Wilson, 1992). The reasonable assumption is that an increase in DO would result in an increase in aerobic biodegradation.

A pilot test was performed at a North Carolina pipeline spill site to confirm the efficacy of air sparging for aeration of an aquifer containing adsorbed and dissolved petroleum hydrocarbons. The lithology at the site is saprolitic in nature with a high degree of heterogeneity. The groundwater table was located approximately 30 feet below the ground surface.

An air sparge well was installed at the site with a screened interval between 55 and 60 feet below the ground surface (25 to 30 feet below water). Three observation wells were installed in a nested configuration at 10 and 50 feet from the sparge point with discrete screened intervals between 15 to 35, 35 to 40, and 45 to 50 feet below the ground surface. In addition, one standard monitoring well was located approximately 95 feet from the sparge well and was used as a control during the test.

Baseline conditions were established prior to initiation of the sparge test. Static DO concentrations in groundwater, groundwater elevation, VOC concentrations in the vadose zone and vadose zone atmospheric pressure were recorded. Compressed air was introduced into the sparge well at approximately 21 to 24 psi and 20 to 42 cfm for approximately 50 hours. All test parameters were monitored on an hourly basis for the first 6 hours and then on a 2- to 12-hour basis for the remainder of the test.

Initial and final DO concentrations in the nested observation wells and monitoring wells are presented in Table 2. Figure 2 shows the increase in DO as a function of time and distance from the sparge point.

TABLE 2. Dissolved oxygen concentrations.

Distance From Sparge (feet)	Screened Interval (feet)	Initial Dissolved Oxygen (ppm)	Final Dissolved Oxygen (ppm)
10	15-35	8.0	10.0
10	35-40	9.0	10.0
10	45-50	0.8	10.0
50	15-35	0.4	1.2
50	35-40	0.4	5.6
50	45-50	1.0	1.2
95	15-35	0.4	0.6

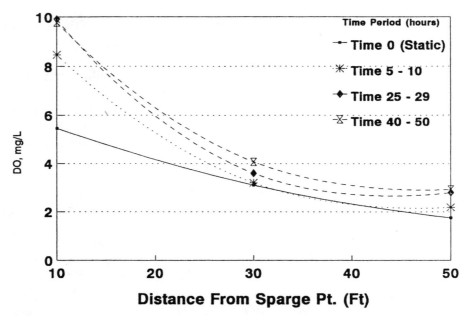

FIGURE 2. Average DO as a function of time and distance from sparge point.

Figures 3, 4, and 5 show the change in DO concentrations in groundwater over the test period in nested monitoring wells located 10 and 50 feet from the sparge point and screened at 15 to 35 feet, 35 to 40 feet, and 45 to 50 feet, respectively. As shown in Figure 2, DO was increased in the aquifer at a distance of at least 50 feet. Saturation (10 ppm DO) was reached at the nearest well (10 ft) after 25 hours. DO levels continued to increase at the further distances over the period of the test. Figures 3, 4, and 5, which portray DO levels at different depths, show that the oxygenation of the aquifer was not uniform. There was a greater increase in DO with distance at the 35 to 40 foot interval than at any other depth. These variable results are likely due to both lateral and vertical heterogeneity in the soil matrix. The DO levels in the nested wells did, however, continue to increase with time indicating vertical mixing which enhances DO distribution.

The results of the pilot test demonstrate both the efficacy and limitations of air sparging for oxygenating an aquifer. DO levels can be increased rapidly at great distances and depths. However, DO response in an aquifer is highly controlled by the lithology. In heterogenous soils, channeling can occur, but is mitigated by vertical mixing through the aquifer.

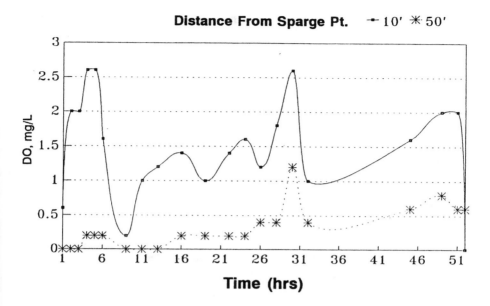

FIGURE 3. Change in DO concentration at 15' – 35' screened interval.

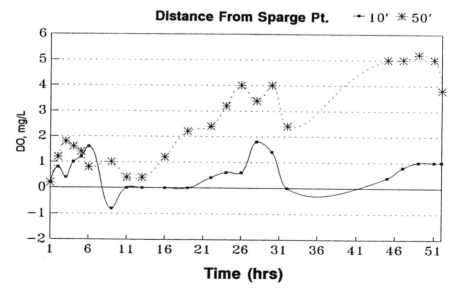

FIGURE 4. Change in DO concentration at 35' – 40' screened interval.

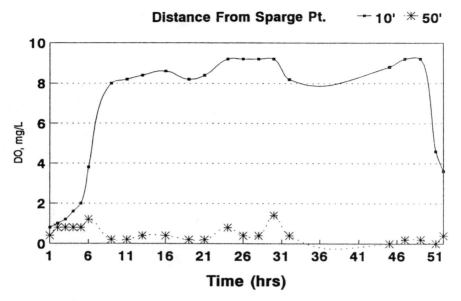

FIGURE 5. Change in DO concentration at 45′ – 50′ screened interval.

LIMITATIONS OF AIR SPARGING

The limitations of air sparging are centered on control factors. Unlike processes based on extraction, such as SVE, that begin with the system under control because contaminants are being drawn to a point of collection, injection systems start with no control because flow is away from the injection point. Control must be gained and maintained. Anything that affects the control of the airflow can limit the application of air sparging. These can be structural or operational.

Air sparging is the controlled injection of air into a saturated soil matrix. The airflow may be impeded by lithological barriers which block the vertical flow of the air or by channelization where the horizontal airflow is "captured" by high permeability channels.

A primary operational concern is overpressurizing the sparge system. This can displace both vapors and water and aggravate the spread of contaminated vapors, NAPLs, or dissolved contaminants. A graphic illustration of the potential displacement of water with injected air is afforded by observing air rotary drilling. Air rotary drilling uses high pressure (~300 to 600 kg/cm^2) and high volume air (8,000 to 12,000 L/min) to lift cuttings. This flow will also cause water to geyser from wells adjacent to the drilling operation.

Overpressurization appears to occur when the injection pressure exceeds ~3 times the breakout pressure. The breakout pressure is the minimum injection pressure required to overcome the water column, i.e., 1 kg/cm² (0.34 psi) for every 0.24 meter of hydraulic head. At pressures above the breakout pressure, air is "injected" laterally into the aquifer. Figure 6 shows that at injection pressures in excess of 3 times the breakout pressure there is no longer a linear relationship between the increase in pressure and the ratio of horizontal to vertical air travel. At low pressures the degree of horizontal travel increases as the sparge pressure increases. A point of inflection occurs at pressures >3 times the breakout, where the increase in injection pressure does not give a corresponding increase in airflow radius. At these high pressures the airflow becomes channelized and water is displaced within the channel.

Air sparging can also spread contamination through accelerated vapor travel. As air sparging increases pressure in the vadose zone, any exhausted vapors can be drawn into building basements which are generally low pressure areas. In areas with potential vapor receptors, air sparging should be done with a concurrent vent system to capture sparged gases. Alternatively, if the contaminant is biodegradable volatile

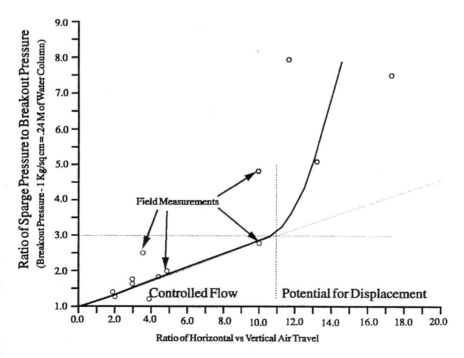

FIGURE 6. Effect of injection pressure on airflow.

removal and the risk of vapor transport may be minimized by adjusting the airflow to maximize biodegradation.

SITE SCREENING

To ensure the effectiveness of a sparge system, proper consideration must be given to the applicability of the technology based on the properties of the contaminant being treated and the geological characteristics of the site. Table 3 summarizes the conditions that are amenable to air sparging. These criteria are not mutually exclusive. Figure 7 shows a logic flow diagram for site selection using these criteria.

SYSTEM APPLICATION AND DESIGN

Proper system design best ensures effective system performance. Air sparging system design requires collection of site data and field pilot testing to identify any subsurface barriers or irregularities which may restrict airflow. The data requirements include the nature and extent of

TABLE 3. Limits to the use of air sparging.

Factor	Parameter	Limit/Desired Range
Contaminant	Volatility	>1 mm Hg
	Solubility	<20,000 mg/L
	Biodegradability	BOD_5[a] >0.01 mg/L, BOD_5:ThOD[b] >0.001
	Strippability	K_H[c] > 10^{-5} atm-m^3-mole^{-1}
Geology	Heterogeneity	No impervious layers above sparge interval. Permeability increases towards grade if layering present.
	Permeability	>10^{-5} if horz:vert is <2:1 >10^{-4} if horz:vert is >3:1
	Saturated Thickness	>5 feet, <30 feet
	Depth to Water	>5 feet

(a) BOD_5 = Biological oxygen demand.
(b) ThOD = Theoretical oxygen demand.
(c) K_H = Henry's law coefficient.

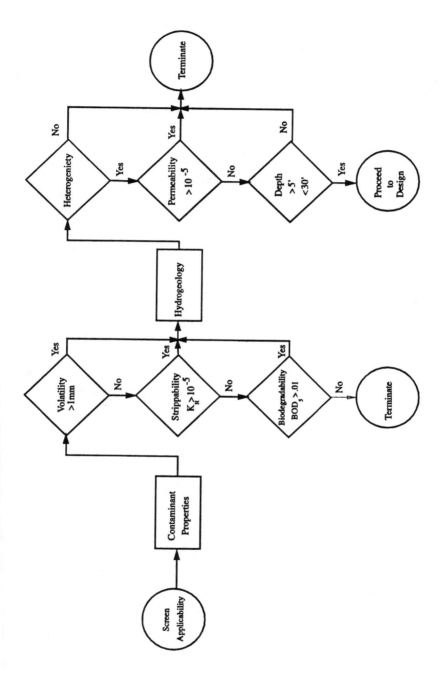

FIGURE 7. Screening procedure for determining the applicability of air sparging.

site contaminants, the site geology/hydrogeology, and thorough knowledge of potential groundwater and vapor receptors. Site contamination
data includes the mass distribution of the contaminants, vertical extent of
adsorbed phase contaminants at or below the water table, lateral extent
of adsorbed phase contamination below the water table, and downgradient
dissolved groundwater concentrations. Geologic parameters include soil
texture, the homogeneity of site soils, and permeability contrasts due to
natural stratigraphic changes or differential filling by human activity.
The proximity of vapor or groundwater receptors would require system
safeguards. Soil vent systems are mandatory where volatile hydrocarbons
are being treated and there are potential receptors, or where vapor phase
controls are required. Groundwater controls such as water collection
systems or a barrier sparge system may be necessary in areas where there
is a concern with dissolved migration.

FIELD PILOT TESTING

Air sparging requires a balanced airflow to maintain remedial
effectiveness and control. Because of the potential for loss of control,
an air sparge system should never be installed without a pilot test. These
design data are determined in pilot testing:

- The radius of influence of the air sparging system conducted
  at different injection flows/pressures
- The radius of influence of the vacuum extraction system (if
  necessitated by the presence of volatile hydrocarbons and
  vapor receptors, or by the need for vapor control)
- The pressure and vacuum requirements for effective treatment
  and effective capture of volatilized materials.

The field tests may consist of a sparge radius of influence test, a
vacuum radius of influence test, and a combined sparge/vent test. The
latter two tests are required where vapor levels are a concern.

A number of different parameters can be measured during the tests
to determine radius of influence, as shown in Table 4. Using multiple
parameters allows for cross correlation during design to determine
effective airflow through the area of contamination and ensure capture
of the volatilized materials. There is generally good agreement between
parameters, as shown in Table 5.

At the conclusion of a properly run site characterization and pilot
test, a complete set of design data will have been collected. Table 6 lists

TABLE 4. Pilot test parameters.

Parameter	Significance
Vacuum/Pressure vs. Distance	Provides an estimate of radius of influence.
VOC Concentrations Soil & Groundwater Static & Dynamic	Indicates area being affected and rates of removal.
CO_2 and O_2 in Soil Vapor	Indicates biological activity.
Increase in DO Levels	Indicates radius of influence. Air travel increases DO.
Water Level Before, During & After	Indicates airflow as air sparging causes water table mounding. Water table "collapses" after sparging.

the data and their significance in design. This data set will ensure that an air sparging system is properly designed, installed, and operated.

FIELD RESULTS

Air sparging is a relatively new technology in general and a very new oxygenation technique for bioremediation in particular. Although there is considerable evidence that air sparging can increase DO levels in aquifers, an important and unanswered question is whether it can enhance bioremediation. Field results of air sparging are presented for a remediation site in Florida as evidence of the efficacy of air sparging.

The aquifer at this site was contaminated with gasoline and diesel fuel compounds. Depth to water was approximately 3 to 5 feet below the ground surface. The lithology at this site was primarily a sandy matrix with a substantial colloidal peat component.

Since 1986, groundwater remediation at this site had been traditional pump-and-treat technology. Dissolved BTEX, naphthalenes, and MTBE concentrations had remained excessively high during the pump-and-treat period. SVE and air sparging was subsequently identified as a supplemental remedial technology. The site was retrofitted with a new recovery well, and an SVE and air sparging systems in 1992.

Air sparging was initiated approximately 20 feet upgradient from this recovery well 14 weeks after groundwater pumping began. Ambient

TABLE 5. Correlation between pilot test parameters in determining radius of influence (ROI).

Site	ROI (m) as determined by:			
	DO	Water Elevation	Pressure	VOC Conc.
Fuel Terminal	1.9	2.7	—	1.8
Fuel Terminal	3.5	3.5	—	3.6
Industrial Site	6.7	6.7	4.9	4.3
Solvent Site	—	13.7	14.2	12.1
Dry Cleaner	—	16.2	18.3	15.3
Pipeline	15.2	—	16.2	—

air was introduced into the aquifer through a sparge well screened from approximately 15 to 20 feet below the ground surface. Air was pumped into the sparge well at a flowrate of approximately 2 to 4 cfm and 4 to 5 psi on a continuous basis.

Figures 8 and 9 show the decrease in concentrations of dissolved benzene, BTEX, MTBE, and naphthalenes during air sparging. The decrease in BTEX concentrations are due to the combined action of stripping and biodegradation. The substantial decrease in naphthalene and MTBE concentrations are particularly indicative of enhanced bio-degradation due to the lower volatility and strippablitiy of these compounds. The decrease in MTBE concentrations are likely due to removal via the pumping system and stripping from the groundwater.

TABLE 6. Site and pilot test data needed for design.

Data	Impact on Design
Lithological Barriers	Feasibility/Sparging Depth
Vertical Extent of Contamination	Sparging Depth
Horizontal Extent of Contamination	Number of Sparge Wells
Volatility of Contaminant	Vapor Control (Venting)
Radius of Influence	Well Spacing/Flow Requirement
Optimal Flowrates	Compressor Size
Vent Radius of Influence (ROI)	Well Spacing
Vacuum/Pressure Balance	Blower Size/Well Placement
Vapor Levels	Vapor Treatment

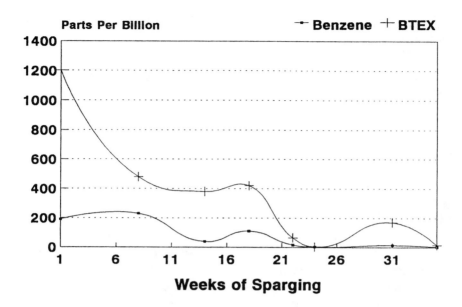

FIGURE 8. Recovery well groundwater quality during air sparging.

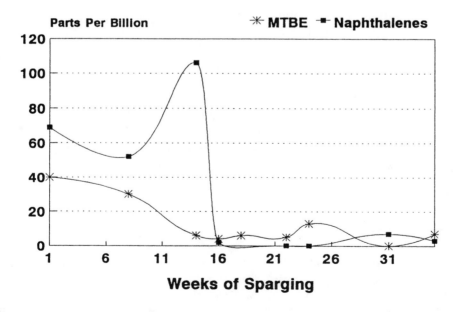

FIGURE 9. Recovery well groundwater quality during air sparging.

These results show the benefits that can be attained with a properly designed, installed and operated air sparging system. The rate of remediation of an aquifer containing strippable and biodegradable compounds can be greatly increased with a SVE and air sparging system. The time required for remediation via groundwater pump and treat alone can be decreased by half with air sparging under favorable conditions.

CONCLUSION

Air sparging is a remedial technology based on the direct injection of air into a water-saturated soil matrix. The injected air travels horizontally and vertically through the soil matrix until it exits into unsaturated (water) soils or directly into the atmosphere. Three potential benefits may be derived from the use of air sparging: volatilization and removal of adsorbed VOCs, in situ stripping of dissolved VOCs from groundwater, and enhanced oxygenation of saturated matrixes.

Of these benefits, enhanced oxygenation is of primary interest in the use of bioremediation. Biodegradation in the saturated zone is limited by the availability of oxygen. When a degradable organic is found in groundwater or in soils below the water table, bacteria will metabolize the organic utilizing the available DO. The groundwater often then becomes anoxic or anaerobic. This is evidenced by the by-products of anaerobic metabolism such as the accumulation of dissolved iron (Fe II) or the presence of partially dechlorinated organics. Additionally, DO measurements in areas of labile contaminants are often below 2 mg/L, as opposed to a typical 5 to 6 mg/L for uncontaminated groundwater. Once depleted, dissolved oxygen is slow to rebound because it must diffuse into the aquifer from the surface of the water table.

Air sparging increases the availability of DO by creating a transient air-filled porosity in the soil pores and by dispersing air throughout the saturated soil matrix. As a result, biodegradation is significantly enhanced.

REFERENCES

Brown, R. A., and J. Crosbie. 1989. "Oxygen Sources for In Situ Bioremediation," Hazardous Materials Control Research Institute, Baltimore, MD, June.

Brown, R. A., J. C. Dey, and W. E. McFarland. 1991. "Integrated Site Remediation Combining Groundwater Treatment, Soil Vapor Extraction, and Bioremediation." In R. E. Hinchee and R. F. Olfenbuttel (Eds.), *In Situ*

Bioreclamation: Application and Investigation for Hydrocarbons and Contaminated Site Remediation, Butterworth-Heinemann, Stoneham, MA.

Brown, R. A., and F. Jasiulewicz. 1992. "Air Sparging Used to Cut Remediation Costs." *Pollution Engineering,* pp. 52-57, July.

Floodgate, G. D. 1973. *The Microbial Degradation of Oil Pollutants.* In D. G. Ahearn and S. P. Meyers, (Eds.) Publ. No. LSU-SG-73-01, Center for Wetland Resources, Louisiana State University, Baton Rouge, LA.

Hiller, D., and H. Gudemann. 1988. "In Situ Remediation of VOC Contaminated Soil and Groundwater by Vapor Extraction and Groundwater Aeration," pp. 2A 90-110. *Haztech International '88,* Cleveland, Ohio, September.

Hutchins, S. R., et al. 1991. *Nitrate for Biorestoration of an Aquifer Contaminated with Jet Fuel,* RSKERL Report, EPA 600/2-91/009, Robert S. Kerr Environmental Research Laboratory, Ada, OK.

Leonard, W., and M. Wilson. 1992. Groundwater Technology, Unpublished results.

Marley, M. C., D. J. Hazebrouck, and M. T. Walsh. 1992. "The Application of In Situ Air Sparging as an Innovative Soils and Ground Water Remediation Technology." *Ground Water Monitoring Review,* pp. 137-144.

Raymond, R. L. 1976. "Beneficial Stimulation of Bacterial Activity in Groundwater Containing Petroleum Hydrocarbons." *AICHE Symposium Series 73,* pp. 390-404.

ZoBell, C. E. 1973. *The Microbial Degradation of Oil Pollutants.* In D. G. Ahearn and S. P. Meyers (Eds.), Publ. No. LSU-SG-73-01, Center for Wetland Resources, Louisiana State University, Baton Rouge, LA.

IN SITU BIOREMEDIATION OF GROUNDWATER CONTAINING HYDROCARBONS, PESTICIDES, OR NITRATE USING VERTICAL CIRCULATION FLOWS (UVB/GZB TECHNIQUE)

B. Herrling, J. Stamm, E. J. Alesi,
G. Bott-Breuning, and S. Diekmann

ABSTRACT

Vertical circulation flow around a specially designed remediation
well (with two screened sections) is discussed. Both physical and
biological in situ groundwater remediations can be carried out
with this technique (UVB/GZB). First, a brief description of the
technique is presented, including a few applications such as an
on-site reactor in combination with the circulation well and/or
the effect of several wells (in one aquifer) on each other. Then
various in situ remediation techniques implemented for different
types of groundwater contaminations are explained; for example,
systems designed for volatile nonaqueous-phase liquids (NAPLs,
e.g., chlorinated solvents) or systems implemented at aromatic
and mineral oil hydrocarbon site remediations (e.g., benzene,
toluene, ethylbenzene, and xylenes [BTEX]). The latter system
not only strips volatile constituents from the groundwater (physical remediation), but also enhances in situ biological contaminant
degradation (bioremediation) by saturating the groundwater with
oxygen. Data of an aromatic and mineral oil hydrocarbon groundwater contamination and remediation verify the successful applicability of this technique in the field. Furthermore, in situ vertical
circulation flow techniques with reactors (e.g., a biofilter) inside
the well casing between the upper and lower screened sections
are discussed. In one pilot-scale test involving pesticides, sorption
as well as biodegradation of the contaminants takes place within

the biofilter. Laboratory results of a complete denitrification process (container filled with special immobile enzymes), that could be used in combination with the vertical circulation flow, are presented.

INTRODUCTION

An increase in the use of human-made organic chemicals over many decades has been observed. Since the middle of this century, the commonly believed concept of "microbial infallibility" (McCarty & Semprini 1992), which means that organic substances can be biologically degraded under any circumstances, has been questioned. Many scientists (e.g., Alexander 1965, Roberts et al. 1989) have conducted research to find degradation pathways for these substances by first determining the limiting factors for biodegradation processes and, second, finding methods to eliminate these factors.

Research on biodegradation processes of groundwater contaminants was successfully carried out in the laboratory; however pilot-scale and full-scale (field) tests implementing the newly developed bioprocesses for in situ groundwater remediations occasionally failed or were not approved by the authorities. Nonetheless, alternatives to pump-and-treat techniques, which have been proven inadequate in many cases, are needed (e.g., U.S. EPA 1989a, 1989b). Due to lower costs of remediation, in situ methods are becoming increasingly popular and should be encouraged more by state officials for various sites and financially for research in the future.

General problems encountered while implementing in situ bioremediations are (1) mixing and evenly distributing nutrients (e.g., N, P), electron acceptors (e.g., O_2), and/or primary substrates (e.g., CH_4) throughout the contaminated aquifer, and (2) too high concentrations of contaminants creating a toxic environment for microorganisms. The vertical circulation flow generated by the UVB/GZB technique (e.g., Herrling et al. 1990, 1991; Herrling & Stamm 1992) can dilute high contaminant concentrations. Originally this technique was used only for in situ remediation of aquifers with strippable contaminants in so-called "vacuum vaporizer wells" (German abbr.: UVB). When no in situ stripping is involved in the process, the technique is called "groundwater circulation well" (German abbr.: GZB). For both the UVB and the GZB, special wells with two screened sections are employed, one at the aquifer bottom and one at the groundwater surface or below an aquitard. A well should be used to remediate only one aquifer (unconfined or confined), and should not connect different aquifers. The groundwater is pumped vertically within the well. The contaminated water enters the well at the bottom and the stripped or treated water leaves at the top or vice versa. In the vicinity of the well a circulation area is created.

Within the well casing, substances can be added to activate biological processes. Nearly constant chemical conditions are reached in the circulation zone (e.g., anaerobic or aerobic). The strong circulation transports the additions to areas partly inaccessible to microorganisms and their enzymes due to an unsuitable environment before the well operation starts.

In addition to discussing the theory behind different techniques, results of a field application, namely a remediation of an aromatic and mineral oil hydrocarbon spill, are presented. Furthermore, promising laboratory results of in situ denitrification are introduced. In the near future this process will be used in containers situated within the well casing. The complete catalytic reduction is realized by immobile enzymes and driven by electrical current.

All technologies described in this paper are patented by B. Bernhardt (see acknowledgment) and S. Diekmann (denitrification) in the USA, Europe, and other countries.

PHYSICAL REMEDIATION
BY IN SITU STRIPPING

The UVB technique has been installed at numerous sites in Germany and more recently in the United States for in situ groundwater remediation of aquifers contaminated with strippable substances (e.g., volatile chlorinated hydrocarbons, or BTEX). As an alternative to conventional hydraulic remediations (pumping, off-site cleaning, and infiltration of groundwater), the UVB technique treats contaminated groundwater by in situ air stripping in a below-atmospheric-pressure field within the remediation well. If contaminants are heavier than water (dense NAPLs or DNAPLs), an upward operating (standard) UVB (Figure 1a) is used, whereas for lighter compounds (light NAPLs or LNAPLs), downward (reverse) flow is implemented (Figure 1b).

The upper, closed part of the well is maintained at below atmospheric pressure by a ventilator. This lifts the water level within the well casing about half a meter. Air for the stripping process is introduced into the system through a fresh air pipe: the upper end is open to the atmosphere, and the lower end terminates in a pinhole plate below the groundwater surface within the well. The height of the pinhole plate is adjusted such

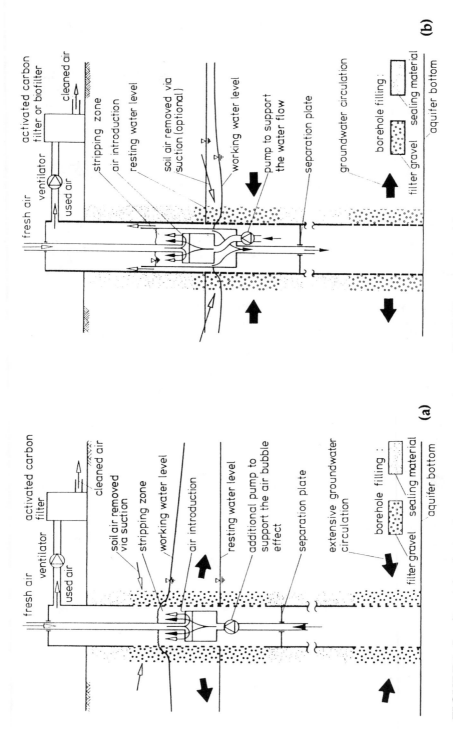

FIGURE 1. In situ stripping using a vacuum vaporizer well (UVB) operating (a) upward for DNAPL (e.g., PCE, TCE), and (b) downward for LNAPL (e.g., BTEX, phenol, kerosene).

that the water pressure is lower there than the atmospheric pressure. Therefore, the fresh air is drawn into the system. The reach between the pinhole plate and the water surface in the well casing is the stripping zone, in which air bubble flow develops. The rising air bubbles produce a pump effect, which moves the water up and causes a suction effect at the well bottom (Figure 1a). In wells used recently, a separation plate and an additional pump (Figure 1) are used to reinforce the pumping effect of the air bubbles. Often the well is simultaneously used for vapor extraction as demonstrated in Figure 1. The off-air charged with contaminants is cleaned by activated carbon or, should suitable contaminants be present, by biofilters. After the ventilator the humidity of the off-air is optimal for sorption of contaminants on activated carbon.

The cleaning effect of the well is based as a result of the air intermixing on the considerable surface area of the air bubbles and on the high concentration gradient between water and clean air. Furthermore, the below atmospheric pressure helps to enhance the escape of volatile contaminants from the liquid to the gas phase. The permanent vibration caused by the air bubbles is beneficial to the escape process of the contaminants. This vibration is transmitted as compression and shear waves into sediment and fluid, and presumably influences the mobility of the contaminants, even outside the well.

The upward-streaming, stripped groundwater leaves the well casing through the upper screen section in the reach of the groundwater surface, which is lifted in an unconfined aquifer by the previously explained pump processes and the below-atmospheric pressure (Figure 1a). A downward operating UVB (Figure 1b) causes a slight depression of the unconfined groundwater surface by the pumping process. The amount is small, however, due to the below-atmospheric pressure, which has the opposite effect. The groundwater then returns in an extensive circulation to the well bottom or to the upper screen, respectively. Thus, the groundwater surrounding the well is remediated. The generated groundwater circulation determines the sphere of influence of a well and is overlapped with the natural groundwater flow.

IN SITU BIOREMEDIATION

When the groundwater is contaminated by other than strippable pollutants, which are suitable for bioreclamation, the extensive circulation flow around the UVB (Figure 2a) can be used to transport nutrients and/or electron acceptors to areas in the saturated zone where they are needed. Any gaseous or liquid substance, soluble in water, can be added in measured quantities while the groundwater passes the well casing. In this

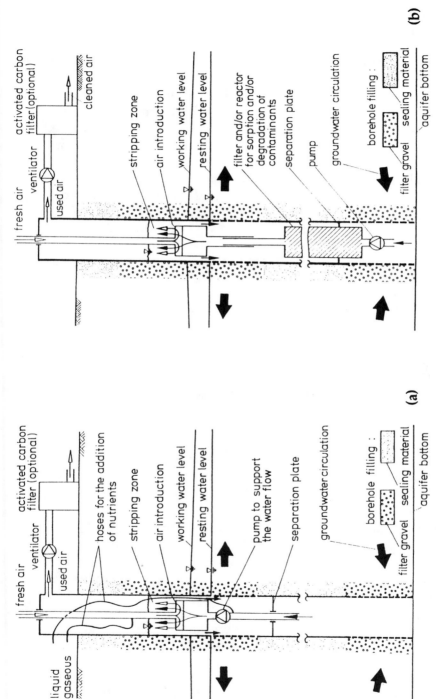

FIGURE 2. In situ bioremediation by a vacuum vaporizer well (UVB) using (a) the aquifer as a bioreactor (e.g., biodegradable hydrocarbons), and (b) special reactors within the well casing (e.g., pesticides, nitrate).

case the aquifer itself is used as a bioreactor. The two-phase flow of gas and water in the stripping zone ensures that the gaseous or liquid addition is most thoroughly mixed with the groundwater. Incomplete mixing could produce undesirable effects in the downstream flow field depending on the respective biodegradation processes.

Essentially, oxygen is needed for in situ bioremediation of, e.g., BTEX or phenol. The in situ stripping process of the UVB provides oxygen saturation of the water. Although the groundwater quantity captured by a well remains relatively constant, the quantity of circulating water around the UVB can be extremely enhanced by a stronger pump in the well casing. Thus the amount of oxygen supplied for in situ bioremediation can be considerably increased with nearly no additional costs. Carbon dioxide as a by-product is removed from the groundwater by the same stripping process, so the pH-value does not drop to produce new problems. Because the hydrocarbons generally are lighter than water, a downward operating UVB is used, in which the contaminated water enters the well through the upper screen (Figure 1b). The contaminated air is cleaned by using e.g. biofilters. A UVB technique that combines vapor extraction with the stripping process generally is used at these sites; thus, large quantities of oxygen are supplied to the vadose zone and capillary fringe for stimulating biodegrading processes. This UVB technique (Figure 1b) combines both physical stripping of volatile organics and biological degradation of pollutants stimulated by the oxygen added as an electron acceptor. If required, nutrients could be added by using the hoses, as demonstrated in Figure 2a. An actual remediation of a BTEX and mineral oil spill is discussed below.

A special bioreactor or other technologies have been or can be installed within the well casing between the lower and upper screens to treat special contaminants (Figure 2b). These in situ techniques can be combined with in situ stripping to eliminate the biogas produced (CO_2 or N_2) or to introduce oxygen.

A specially designed well based on the UVB technology of Figure 2b was used in a pilot study for a biological and physical in situ remediation of triazines. During reloading and rinsing out of the storage tanks, pesticides seeped into soil and groundwater. Downgradient of this source area, where the UVB is located, the contaminated groundwater flows into the remediation well through the lower screen section and is pumped through the activated carbon filter within the well casing and up to the reactor situated aboveground for this pilot study. Here, oxygen saturation takes place by the standard UVB technique (Figure 2b). Activated carbon is used to adsorb the triazines; the carbon also serves as backing and growth material for microorganisms that exist in the groundwater. Before the pilot test began, triazine metabolites were detected in the groundwater. Thus, contaminant-adapted microorganisms were present and already degraded some triazines.

Typical factors limiting the biological degradation rate include low triazine concentrations in the aquifer and low absolute numbers of microorganisms in the groundwater. By allowing both the contaminants and the microorganisms to accumulate within the biofilter, these limitations can be overcome. Sufficient oxygen for the degradation is available because of the oxygen saturation of the groundwater. Thus, the activated carbon unit need not be renewed often. The collected field data and experiences obtained from this pilot test verify that the UVB with a biofilter within the well casing can be successfully applied to triazine contamination in groundwater (Buermann et al. 1992).

Groundwater contaminated by nitrate can be remediated in situ in containers installed within the well casing between the two screen sections of a UVB (Figure 2b). This in situ denitrification technique is discussed below.

ON-SITE REMEDIATION USING A GZB

When an on-site technique for groundwater remediation is needed, e.g., to eliminate dissolved heavy metals from the groundwater, vertical circulation flow of a GZB can be used. Groundwater entering the well is pumped aboveground, treated, and infiltrated in the same well using the other well screen (Figure 3).

Following these ideas, the GZB can be used as a pump or infiltration well for standard on-site remediation. In such a case, partial withdrawal or infiltration is taken from or added to the total volume of groundwater flowing through the well casing (see Figure 2 in Herrling & Stamm 1992). It is possible to extract or to infiltrate water without changing the groundwater head at the well top for a special ratio of the extracted or infiltrated quantity of water and of that vertically circulating around the well. This has been numerically investigated for confined aquifer conditions by Herrling and Stamm (1992). At some distance from the well, the groundwater head at the aquifer top slightly deviates from the resting position. If the well capacity is low, pumping is continuous at a much higher rate and the infiltration can be realized for a much greater quantity, even for a low distance between the surface and the groundwater table. While pumping, the turbulent mixing of air and water in the artificial sand pack of the well construction, which often causes unwanted precipitation, can be avoided.

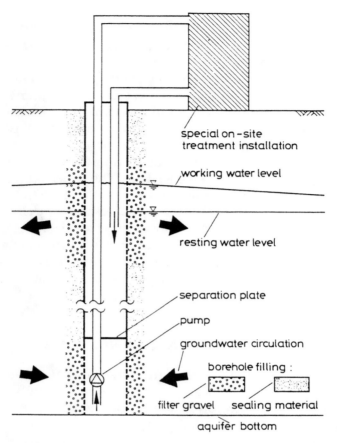

FIGURE 3. On-site treatment using a groundwater circulation well (GZB) to eliminate dissolved heavy metals.

IN SITU TREATMENT WALL GENERATED BY VERTICAL CIRCULATION FLOW

When a UVB or a GZB is situated in a plume of contamination, the polluted upstream groundwater (1) is captured by the well, (2) is treated by suitable in situ (or on-site) measures, (3) circulates vertically in the surrounding of the well, and (4) is released by the well as treated water flowing downstream. For a wide plume, several UVBs or GZBs are arranged in one line normal to the natural groundwater flow. The distance between each system is determined such that no water can pass the well line without having been treated. The natural flow field is only locally influenced because no groundwater is extracted. Thus, a line of several wells represents a treatment wall and no contamination coming from upstream can pass the line of wells.

NUMERICAL INVESTIGATIONS

Numerical investigations of vertical circulation flow around the GZB and UVB are continuing (Herrling & Stamm 1992). Previously, only flow for confined aquifer conditions was studied, which permits the super-position of the flow fields of different wells and the natural groundwater flow; the local below-atmospheric-pressure field (in case of a UVB) was neglected. Further, density effects were ignored; only steady-state conditions were taken into account, and to estimate the capture zone, only convective transport was considered.

The general character of vertical circulation flow is shown in Figure 4. In a vertical longitudinal section parallel to natural groundwater flow, streamlines mark the flow around one (Figure 4a) and two (Figure 4b) upward-pumping UVBs or GZBs, that are separated by the distance between the stagnation point and the well axis. The strong vertical circulation flow especially between the two wells (Figure 4b), which is extremely beneficial in a highly polluted area near the contamination source, is evident. Two wells in a row in the flow direction (Figure 4b) can be used to generate an anaerobic biodegradation zone around the first circulation well (GZB without air stripping, but with carbon source added within the well casing) and an aerobic zone with a UVB around the second well.

A view of the numerically calculated separating stream surfaces of nine water bodies in connection with the flow around three UVB or GZB

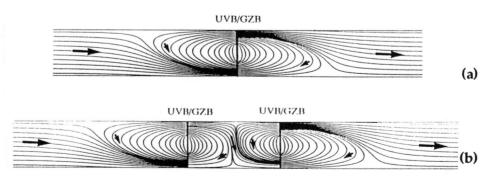

FIGURE 4. Streamlines around (a) one and (b) two UVBs or GZBs demonstrated in a vertical longitudinal section parallel to the natural groundwater flow.

is presented in Figure 5. They are positioned at a maximum distance normal to the natural groundwater flow so that no water can pass between the wells without having been treated. With the contaminated groundwater flowing from the left to the right the following salient features can be seen: the separating stream surfaces of the contaminated water captured by the three UVB or GZB (left), the surfaces of the water bodies having been treated and circulating around the wells (center), and the treated water flowing downstream (right). The numerical investigations resulted in diagrams for the dimensioning of UVB or GZB installations (Herrling et al. 1990, 1991; Herrling & Stamm 1992).

UVB REMEDIATION OF AROMATIC HYDROCARBONS (MAINLY BTEX) AND MINERAL OIL GROUNDWATER CONTAMINATION

Case Description

The UVB technology combining in situ stripping and in situ bioremediation (Figure 1b) was installed in Berlin, Germany, and is successfully remediating a site contaminated with BTEX, styrene, and mineral oil hydrocarbons. The spill was caused by a leaking storage tank. Some feasibly removable soil was excavated from the unsaturated zone and transported to a landfill. The concentration of BTEX and styrene in the groundwater at the beginning of remediation was 280 mg/L and of mineral oil hydrocarbons was 1 mg/L. An assumed contamination of chlorinated hydrocarbons was of less significance.

Geologic and Hydrogeologic Situation

Underneath approximately 1.2 m fill material consisting of fine sand and fragmented bricks lie 2.7 m of medium to fine sand with recognizable graded bedding. From 3.9 m to about 10.6 m below the soil surface sandy gravel dominates; however, clayey silt as well as fine organic-rich layers and occasional stones form intermittent horizontal layers. At around 10.6 m, gray clayey marls make up an impermeable layer.

The unconfined aquifer has a thickness of about 6.5 m; the groundwater table lies at around 4 m below the ground surface. With a hydraulic gradient of 0.08% and a hydraulic conductivity of 3.1×10^{-3} m/sec,

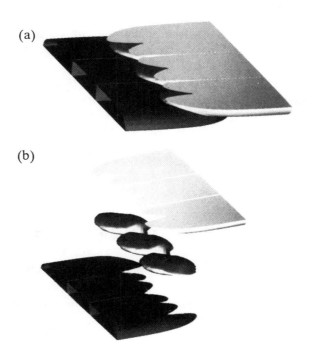

FIGURE 5. Separating stream surfaces of the different water bodies in the outside flow of a UVB or a GZB: captured, circulating, and flowing downstream water in (a) a real situation, and (b) water bodies separated for clarification.

groundwater flows at a rate of 1.1 m/day to the southwest. Pore space in the sandy gravel is 0.2.

Remediation Measures

Two UVBs were installed to remediate the groundwater and vadose zone at the site (site map see Figure 6). The source area of the contamination was just upstream of UVB 1. Both UVB 1 and UVB 2 were drilled with a borehole diameter of 0.8 m, have a well diameter of 0.4 m, and are 10.9 and 11.2 m deep, respectively. Because free product of LNAPL was found on the groundwater surface, both UVBs operate downward (see Figure 1b). The well construction of UVB 1 is shown in Figure 7; that of UVB 2 is similar. The screened sections are from 3.1 to 5.3 m and 9.0 to 10.5 m (UVB 1) and from 3.15 to 5.35 m and 8.2 to 10.2 m (UVB 2) below ground surface.

The remediation started with a 5-day test run of UVB 1 in October 1990; about 88 kg contaminants were removed via the off-gas of the stripping process. After a 4-month stoppage by the client, UVB 1 was turned on again in March 1991. Technical modifications were carried out between March 29 and June 19, 1991. The operation of UVB 2, which is located 40 m downstream of UVB 1, started on October 22, 1991.

Modification of UVB 1 included placing the stripping reactor directly above the well as opposed to within the well. This enabled a better control of the biomass production. UVB 2 has the same design.

Results of the UVB Remediation

Approximately 164 kg aromatic hydrocarbons (AHCs) and mineral oil hydrocarbons (MHCs) had been removed from the groundwater and vadose zone via off-air of UVB 1 through May 1992. Figure 8 shows the

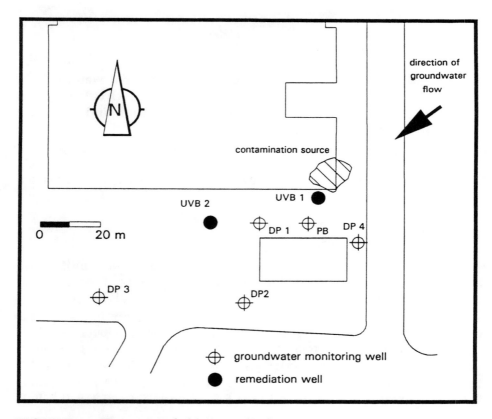

FIGURE 6. Site map of the remediation.

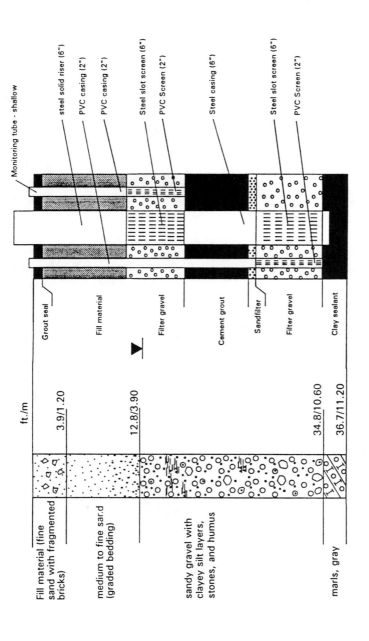

Geologic profile   Schematic well construction

FIGURE 7. Geologic profile and schematic well construction of UVB 1.

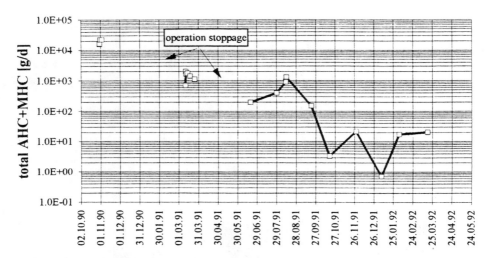

FIGURE 8. Daily contaminant rate in the off-gas of UVB 1.

daily total contaminant rate. The AHC concentrations of the influent and effluent groundwater of the UVB 1 system are presented in Figure 9.

Approximately 10 g of contaminants were removed from UVB 2 through May 1992, most of which was extracted within the first 2 weeks of operation. Contrary to expectations, the contaminant concentrations in the groundwater entering UVB 2 remained below detection limits during

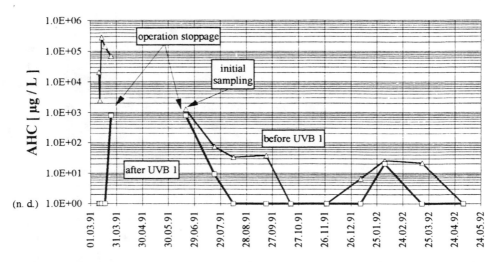

FIGURE 9. AHC concentrations of the influent and effluent ground-water of the UVB 1 system.

the operation. The circulation zone around UVB 2 is fully saturated with oxygen as a consequence of the stripping process (see Figure 10b). Thus, a percentage of the contaminants could be biodegraded within the aquifer before the groundwater reaches the well. We assume this because concentrations of 3 mg/L were analyzed for groundwater from monitoring well DP 1 (15 m upstream of UVB 2, see Figure 6) in October 1991. This concentration dropped to below the detection limit within 30 days.

Biological processes most likely also have been responsible for contaminant reduction in the groundwater around UVB 1. Biomass coating with a thickness of about 2 cm was found on the inside of pipes and the casing of the stripping reactor. The reactor itself was covered with a thin biocoating. Furthermore, within the well casing a considerable amount of biomass was found. Nevertheless the pump rate of 6 m³/h could be kept up without problems during the well operation.

Oxygen concentrations and pH values of the groundwater circulation around UVB 1 were determined (Figures 10 and 11). Before remediation began and when the UVBs were not in operation, nearly anaerobic conditions and pH values of about 8.2 were found in the groundwater. When the UVB system was running, however, the stripping reactor saturated the groundwater with oxygen, achieving concentrations of nearly 10 mg/L. The groundwater leaving the UVB 1 well through the lower screen section (controlled by the deep monitoring tube) had a much lower oxygen content than expected (see Figure 10a), which indicates that biodegradation occurs. In contrast, the oxygen levels measured in the deep monitoring tube of UVB 2, where a lack of contaminants exists, remain at approximately O_2 saturated levels suggesting no oxygen consumption by biological processes (see Figure 10b). Biodegradation takes place not only within the well, but also in the circulation zone supplied with oxygen. Thus, the groundwater saturated zone can be regarded as a bioreactor (see Figure 10a). One should keep in mind that in addition to contaminant breakdown, other processes in this zone could consume oxygen. Furthermore, approximately 30% of nearly anoxic upstream groundwater is always entering the circulation zone.

Figures 10a and 10b show that the oxygen consumption within the circulation zones of both UVB increases with remediation time. The increasing difference of the O_2 concentrations in the influent and effluent groundwater suggests an increase in biodegradation in spite of the decrease of contaminant concentration in the groundwater (see Figure 9).

It is important to note that, as the groundwater passes through the stripping reactor of UVB 1, its pH rises (see Figure 11a); this means the produced biogas CO_2 is stripped out of the groundwater. Because

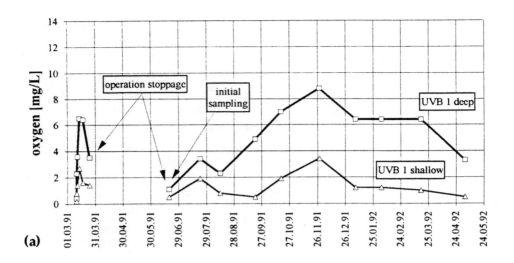

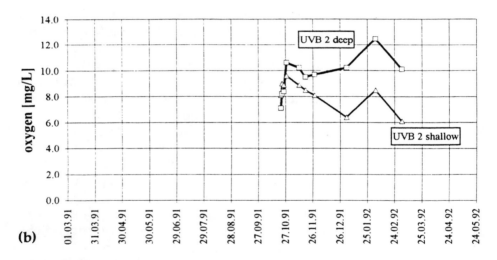

FIGURE 10. O$_2$ concentrations in the shallow and deep monitoring tube of (a) UVB 1 and (b) UVB 2.

biodegradation is much lower around UVB 2, this phenomenon does not occur to such a degree (see Figure 11b). Further observations also suggest that biogas is produced in the aquifer. For example, at monitoring well DP 1, 25 m downstream of UVB 1, the pH value decreased to about 7.3 after January 1992. Nearly no oxygen can be measured at DP 1, although the effluent of UVB 1 delivers an increasing amount.

IN SITU DENITRIFICATION

The high concentration of nitrate in many groundwaters is a serious environmental problem. Historically, nitrate has been removed from drinking water using physiochemical techniques such as electrodialysis, reverse osmosis, and ion exchange. The one factor these techniques have in common is that the nitrate is removed from the original solution but

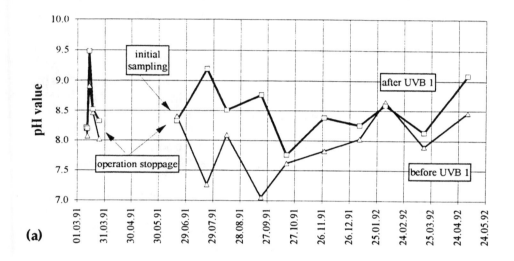

(a)

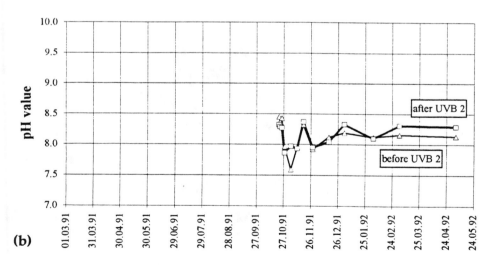

(b)

FIGURE 11. pH value before and after the stripping unit of (a) UVB 1 and (b) UVB 2.

is not degraded and is given back to the environment in one way or another in much higher concentrations. Alternatively, nitrate can be directly reduced to nitrogen gas using biological methods. A variety of biological-bacterial reactions are commercially available, many of which use mixed bacterial cultures or immobilized denitrifying bacteria. Such bioreactors are delicate to use, because the bacteria require the addition of C and P sources to the water before purification and they typically need a longer run up time. The three enzymes needed for technical denitrification (nitrate, nitrite, and N_2O reductase) exist in only small amounts in the bacterial cells.

By immobilizing these enzymes with artificial cofactors and electrochemical reductant, a denitrifying matrix can be constructed with a specific catalytic activity almost 3 orders of magnitude greater than that using immobilized bacteria. The matrix with the three immobilized reductases (and coimmobilized cofactors) is packed in a flow-through chamber, and the reducing agents are offered by an electrode. Figure 12 depicts an electrical column (electrodes 9.5 cm × 4.5 cm conductive plastic plate, KS 2727-EL [BASF]) with 2 mL nitrite reductase/N_2O reductase/Safranin T matrix (2 units enzyme) packed within 1 mm of the cathode. Arrows indicate the direction of water flow. The dashed line indicates the position of a 90-μm pore-size glass filter. The drinking water is purified upon passage through the matrix chamber. This nitrate-degrading process can be combined with the stripping system presented above (Figure 2b).

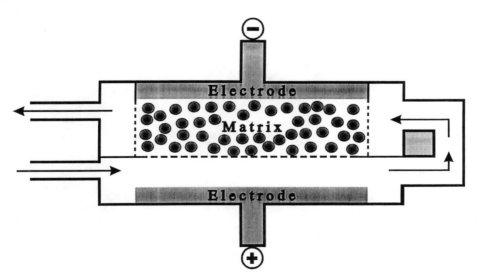

FIGURE 12. Electrical column packed with a denitrifying matrix in a flow-through chamber.

Immobilization of these three reductases (nitrate, nitrite, and N_2O) is particularly difficult because it involves the coimmobilization of electron-transferring cofactors. The physiological electron donors (for example, reduced NAD [nicotinamide-adenine-dinucleotide] or reduced cytochrome) are not effective under these conditions. Instead, alternative electron donors including dyes of the viologen family are used (Mellor et al. 1992). Table 1 presents measured results. Nitrate and nitrite reductase activity was tested with two immobilized dyes as cofactors as percent of methyl viologen (enzyme activity [0.1 unit added]). The given numbers are the mean of four independent repeats; the error is estimated to be ±10% (free), free enzyme and immobilized dye, or (immob.) dye and enzyme coimmobilized. The unit is defined in this way: 1 unit nitrate reductase produces 1 µ mole nitrite per minute; 1 unit nitrite reductase removes 1 µ mole nitrite per minute. One unit nitrous oxide reductase produces 1 µ mole nitrogen gas per minute. Nitrate reductase was produced from *Zea mays* and nitrite reductase from *Rhodopseudomonas*.

To avoid water contamination, no chemical compound should be added to drinking water to provide the final reductant of the enzyme and cofactor. Therefore, the reducing properties of electrodes (cathodes) were used as arranged in Figure 12. Using this system, both nitrate and nitrite substrate turnover could be carried out.

Figure 13 presents the nitrate (•) or nitrite (+) turnover in an electrical column as in Figure 12 (3 V, packed within one unit either nitrate reductase/Azure A or nitrite reductase/Safranin T matrix) versus flowrate expressed either as turnover per milliliter (Figure 13a) or total efficiency (Figure 13b). The substrate concentration is 1 mM.

When such column compartments containing different active matrices were joined in series (nitrate reductase followed by nitrite/N_2O reductase), significant nitrate degradation could be demonstrated. Table 2 shows the reduction of nitrate by coupled electrical columns of two nitrate concentrations (I is 1.0 mM and II is 10 mM). The nitrate solutions were

TABLE 1. Test results of the used electron donors.

Reductases	Nitrate		Nitrite	
Enzyme source	*Zea mays*		*Rhodopseudomonas*	
	free	immobilized	free	immobilized
Azure A	30.8	97.6	2.7	62.5
Safranin T	44.5	75.0	55.7	104.1

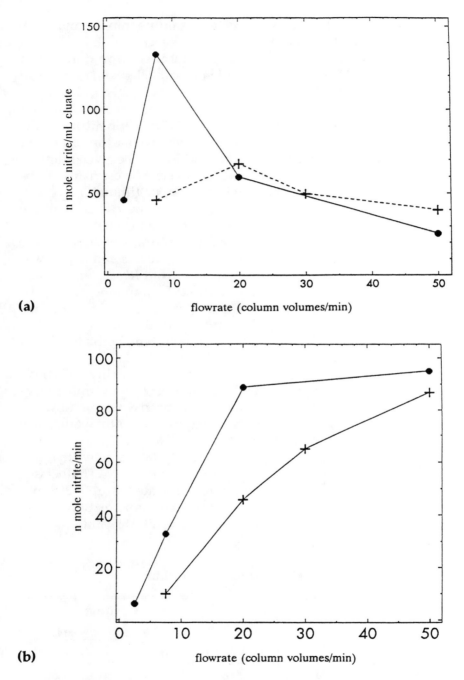

(a)

(b)

FIGURE 13. Nitrate (•) and nitrite (+) turnover in an electrical column versus flowrate (a) turnover per milliliter and (b) total efficiency.

TABLE 2. Nitrate reduction by coupled electrical columns starting from different nitrate concentrations.

	Substrate NO_3 Concentration (mM)	Gas Evolution (μmole/min)		
	NO_3	NO_2	N_2O	N_2
I				
substrate	1.0	0	0	0
eluate	0	0	0	2.5
II				
substrate	10.00	0	0	0
eluate	8.69	0.38	0	2.3

passed through a column with two compartments contained in an airtight, helium-filled chamber. The first compartment contained 5 units nitrate reductase coimmobilized with Azure A. The eluate from this compartment was fed into a second compartment containing 8 units nitrite reductase/N_2O reductase enzyme mix coimmobilized with Safranin T. The flowrate was 5 compartment volumes per minute (5 mL/min). The potential difference was 4 V. Nitrate was estimated enzymatically with reference to a standard curve. Substrate concentrations remained unchanged after passing through the column compartments without electrical voltage applied. Gas eluate was measured between 20 and 30 min after applying the voltage. The errors in the salt measurements are less than ± 1%, in the gas measurements less than ± 6%. The limit of detection for N_2O lies under 0.1 μmol. Calculations of the N balance are done with nitrogen measured in the gas phase as N_2.

Concomitant nitrogen production was observed in molar equivalent ratios to the substrate added. From the results displayed in Figure 13, we determined the specific catalytic activity of the columns to be at least 560 kg NO_3 per m^3 matrix and day. Similar results were obtained using alternative matrices (glass particles) and immobilization chemistries used in food technology. However, it should be noted that our results are obtained with small reactors in our laboratories. Currently this technology is in the technical developmental stage.

For commercial use, stability and enzyme price are important factors. Figure 14 shows that enzyme activity remained in the column for at least 3 months. Calculations predict that this process will become practical

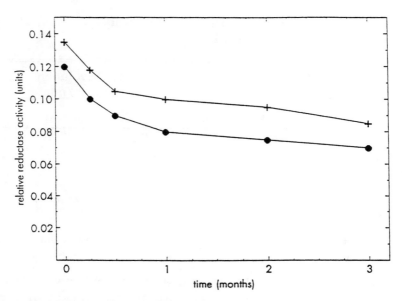

FIGURE 14. Measured enzyme activity of nitrate reductase/Azure A (+) and nitrite reductase/Safranin T (•).

at enzyme prices of US $ 3 per unit of enzyme, resulting in a price of US $ 0.16/m³ of processed water. Figure 14 demonstrates the stability of the nitrate reductase/Azure A (+) and nitrite reductase/N_2O reductase/Safranin T (•) matrices kept at 4°C and periodically assayed; 0.1 unit methyl viologen enzyme activity originally was immobilized.

In principle, the applications of the process presented here are not limited to nitrate reduction. With the appropriate enzyme, any substrate could be manipulated such as phosphate; sulfate; and aryl, alkyl, or aromatic halides (for example pesticides). Oxidoreductases represent the largest group of enzymes known, indicating that a large range of substrates can, in principle, be degraded.

CONCLUSION

These technologies, in particular vertical circulation flow around wells with two screen sections, offer many advantages, especially compared to the standard pump-and-treat system (Herrling et al. 1991, who discuss bioremediation in connection with circulation flow). We have focused on two techniques. The first one uses the aquifer as a bioreactor and combines in situ bioremediation with in situ stripping (case study on aromatic and mineral oil hydrocarbons), and the second one employs special bioreactors within the well casing and uses the stripping process for its oxygen supply (pesticides) or for removing biogas (nitrate).

The case study has demonstrated that AHCs and MHCs can be successfully stripped and biodegraded at the same time. The lack of financial resources during a commercial remediation and the complexity of a field application allow neither a precise balance nor a more detailed investigation concerning the fate of all the contaminants. Some microbiological investigations were carried out (biodegradation tests in the laboratory) but are, unfortunately, incomplete for lack of financial resources. At the biological processes, the following could be ascertained:

1. The production of biomass and biogas (CO_2) was detected.
2. Bioclogging was not a limiting factor of the bioremediation.
3. Oxygen consumption in the aquifer increases with time, although the measured contaminant concentrations at the well decrease, indicating that the circulation zone within the aquifer increasingly becomes a bioreactor.
4. The bioprocess can be enhanced by increasing the well discharge with a stronger pump.
5. A second UVB downstream can help strip out the biogas and lift pH values, besides offering additional remediation.
6. The contaminant concentrations in the groundwater (originally 280 mg/L) decrease within a short time (< 1.5 years) to nondetectable values.

Sorption of pesticides on activated carbon in combination with their biodegradation as well as complete denitrification with immobile enzymes, where the reduction is driven by electrical current, are two promising techniques, carried out in containers within the well casing, to clean groundwater in situ from those contaminants.

ACKNOWLEDGMENTS

The first two authors gratefully acknowledge IEG mbH, D-7410 Reutlingen (Germany), for financially supporting these investigations. In particular we thank B. Bernhardt, IEG mbH, D-7410 Reutlingen, inventor and patent holder of the UVB and GZB method. We also thank P. Brinnel, PROTEC GmbH, D-6370 Oberursel; W. Buermann, University of Karlsruhe, D-7500 Karlsruhe; W. Kaess, D-7801 Umkirch; and H. J. Lochte, UTB mbH, D-4020 Mettmann, for many helpful discussions and contributions.

REFERENCES

Alexander, M. 1965. "Biodegradation: Problems of Molecular Recalcitrance and Microbial Fallibility." In *Advances in Applied Microbiology*, pp. 35-80. Academic Press, New York.

Buermann, W., G. Bott-Breuning, and R. Krug. 1992. "Groundwater Remediation Using the Vacuum-Vaporizer-Well." In W. Pillmann (Ed.), *Industrial Waste Management (Proc. Envirotech Vienna 1992)*, pp. 723-732. A. Riegelnik Printers, Vienna, Austria.

Herrling, B., W. Buermann, J. Stamm, and M. Schoen. 1990. "UVB Technique for In-Situ Groundwater Remediation of Strippable Contaminants: Operation and Dimensioning of Wells." In W. Pillmann and K. Zirn (Eds.), *Hazardous Waste Management, Contaminated Sites and Industrial Risk Assessment*, pp. 631-640. A. Riegelnik Printers, Vienna, Austria.

Herrling, B., W. Buermann, and J. Stamm. 1991. "Hydraulic Circulation System for In Situ Bioreclamation and/or In Situ Remediation of Strippable Contamination." In R. E. Hinchee and R. F. Olfenbuttel (Eds.), *In Situ Bioreclamation, Applications and Investigations for Hydrocarbon and Contaminated Site Remediation*, pp. 173-195. Butterworth-Heinemann, Boston, MA.

Herrling, B., and J. Stamm. 1992. "Numerical Results of Calculated 3D Vertical Circulation Flows Around Wells with Two Screen Sections for In Situ Aquifer Remediation." In T.F. Russel et al. (Eds.), *Computational Methods in Water Resources IX*, Vol.1: *Numerical Methods in Water Resources*, pp. 483-492. Elsevier Applied Science, London.

McCarty, P. L., and L. Semprini. 1992. "Engineering and Hydrological Problems Associated with In Situ Treatment." In S. Lesage (Ed.), *Proc. In-Situ Bioremediation Symposium '92*, pp. 2-13. Wastewater Technology Centre, Burlington, Ontario, Canada.

Mellor, R.B., J. Ronnenberg, W.H. Campbell, and S. Diekmann. 1992. "Reduction of Nitrate and Nitrite in Water by Immobilized Enzymes." *Nature 355*: 717-719.

Roberts, P. V., L. Semprini, G. Hopkins, D. Grbić-Galić, P. L. McCarty, and M. Reinhard. 1989. *In-Situ Aquifer Restoration of Chlorinated Aliphatics by Methanotrophic Bacteria*. U.S. Environmental Protection Agency, EPA/600/2-89/033, Cincinnati, Ohio.

U.S. Environmental Protection Agency. 1989a. *Evaluation of Ground Water Extraction Remedies*, Vol. 1, *Summary Report*, EPA/540/2-89/054, Cincinnati, Ohio.

U.S. Environmental Protection Agency. 1989b. *Evaluation of Ground Water Extraction Remedies*, Vol. 2, *Case Studies 1-19*, EPA/540/2-89/054a, Cincinnati, Ohio.

THE DESIGN AND MANAGEMENT OF SYSTEM COMPONENTS FOR IN SITU METHANOTROPHIC BIOREMEDIATION OF CHLORINATED HYDROCARBONS AT THE SAVANNAH RIVER SITE

K. H. Lombard, J. W. Borthen, and T. C. Hazen

ABSTRACT

The successful operation of an in situ bioremediation system is inherent within its design. Well-organized system components enable ease of maintenance, limited down time, and relatively rapid data acquisition. The design effort in this project focused on injection of a low-pressure air/methane mixture into a horizontal well below the water table, a methane-blending system that provided control of the injected mixture, redundant safety interlocks, vapor-phase extraction from a second horizontal well, and an off-gas treatment system that provided efficient thermal catalytic oxidation of the extracted contaminant vapors. The control instrumentation provided sufficient redundancies to allow the system to remain in operation in the event of a component failure, and equally important, the safe shutdown of the system should any designed safety parameters be exceeded (i.e., high methane concentration). Final design approval took into consideration the reliability of the equipment and the components specified. Product knowledge and proper application limited the risk of a component or system failure while providing a safe, efficient, and cost-effective remediation system. Microprocessor data acquisition and system control were integrated with an auto-dialer to provide 24-hour emergency response and operation without on-site supervision. This integrated system also insured accurate data analysis and minimum downtime. Since operations commenced, the system has operated a total of 7,760 hours out

1-56670-084-1/94/$0.00 + $.50

of the possible 8,837 hours available. This equates to an operating
efficiency of 87.8%.

PURPOSE

A recent search of the literature regarding in situ bioremediation
revealed only limited sources of information associated with the design
and engineering of the surface supply and support systems. A successful
biostimulation of the indigenous microbial community in a subsurface
environment depends on effective delivery of key components (i.e.,
nutrients and electron acceptors) and proper monitoring to ensure safe
and consistent delivery.

It is the intent of this paper to provide the reader with additional
insight through lessons learned and, to provide the knowledge necessary
to aid in the design and construction of an effective in situ methanotrophic
bioremediation system. It is our objective to provide continued research
and development of new and innovative engineering concepts so that
safe, efficient, and cost-effective systems can be realized.

INTRODUCTION

Site History

Contamination of the Savannah River Site's M-Area ground (vadose
zone) and groundwater by chlorinated solvents from M-Area degreasing
facilities evolved over a 35-year period. It is estimated that from 1952
to 1982, M-Area used approximately 13,000,000 lb (5,900,000 kg) of tri-
chloroethylene (TCE) and tetrachloroethylene (PCE) as degreasing solvents.
The degreasing processes resulted in the evaporation of 50 to 95% of the
solvents. An estimated 2,000,000 lb (947,200 kg) of TCE and PCE may
have been released to the M-Area process sewer system leading to the
M-Area settling basin and some 1,500,000 lb (710,400 kg) to the A-014
outfall (Marine & Bledsoe 1984).

The following abbreviated history of the site's construction efforts
and solvent usage is adapted from Marine and Bledsoe (1984) and recast
by Looney and Rossabi (1992). This information will help the reader
understand the magnitude of the problem that exists, and the engineering
challenges faced today, in the research and development of new technologies for cleanup operations of hazardous waste sites.

To support the degreasing operations, a terra-cotta process sewer line
and an 8,000,000-gal (30,280,000-L) unlined surface impoundment, used

for a settling basin, were constructed. Effluent disposal to seepage or settling basins was standard industrial practice during the 1950s through the 1970s.

Dissolved solvents were identified in the groundwater beneath the settling basin in 1981. A subsequent investigation of the process sewer line revealed cracks and plant roots penetrating the terra-cotta pipe. In 1984 the pipeline to the basin was relined and by 1985, the process wastes from M-Area were diverted to the Liquid Effluent Treatment Facility, and settling basin use was discontinued.

The Savannah River Site (SRS) voluntarily commenced groundwater remediation with a full-scale pump-and-treat system in April of 1985. In the operating permit from the South Carolina Department of Health and Environmental Control (SCDHEC), the SRS made a series of commitments, one of which was to develop and evaluate new technologies to improve system performance. To date, several innovative projects have successfully demonstrated the SRS's commitment to the environmental restoration of the M-Area settling basin system. As the host site for the U.S. Department of Energy's Integrated Demonstration Program, the SRS is conducting a multiphased experiment, implemented to test new remediation and monitoring technologies and innovative engineering concepts along the abandoned process sewer line. Part of this integrated demonstration includes the In Situ Bioremediation Demonstration as part of the cleanup of organics from sediments and groundwater at non-arid sites.

THE IN SITU
BIOREMEDIATION DEMONSTRATION

Phase I In Situ Air Stripping and Geophysical Monitoring

Phase I of the project included extensive site characterization. The geology of the site consists predominantly of unconsolidated sands and clayey sands with interbedded clay horizons. These tertiary sediments were deposited in shallow marine, lagoonal, and fluvial environments (Eddy et al. 1991). Many geophysical devices were deployed, and water table monitoring wells were installed to provide data to help determine the effectiveness of using air stripping in horizontal wells to remediate a contaminated site.

In 1989, two horizontal wells were installed along the abandoned sewer, in what was considered to be a hot spot, an area of high concentrations of contaminants (TCE and PCE). During the experiment, air was injected into the horizontal well below the water table, at varying rates of 65, 170, and 270 scfm (31, 80, and 127 L/sec) and extracted from the upper horizontal well located in the vadose zone, at a constant rate of 580 scfm (274 L/sec). During the 139-day test, nearly 16,000 lb (7,258 kg) of volatile organic compounds were removed from the subsurface (Looney et al. 1991).

The information from this experiment established the baseline data for the Phase II Methane Injection Campaign, which is intended to stimulate the methanotrophic bioremediation of the volatile organic contaminants.

Phase II Methane Injection Campaign

The goals of the In Situ Bioremediation Demonstration and design criteria for the Methane Injection Campaign were established by an expert panel of scientists and engineers using Phase I data retrieved during the in situ air stripping operations. A detailed description of the project is documented in the Test Plan (Hazen 1991). The air/methane injection rate was set at 200 scfm (94 L/sec). In order to provide the required air flow, a compressor capable of delivering 300 scfm (142 L/sec) at 100 psig (45 kg/cm^2) was specified.

An extraction rate of 20% greater than the injection rate was established. This strategy was adapted to prevent plume expansion while stimulating the bacteria in the saturated sediments and vadose zone, and to prevent the methane from reaching a potentially explosive concentration of 5%, the lower explosive limit (LEL). To provide the necessary vapor extraction rate, a vacuum-blower unit capable of generating 500 scfm (236 L/sec) at 10 in. (25.4 cm) Hg (inlet), was specified.

The final design parameter for the extracted gases, was to reduce the total organic contaminants in the off-gas to a concentration of less than 5 ppm (vol/vol). To facilitate this an Allied Signal Halocarbon Destructive Catalytic (HDC) oxidation system, Air Resources Inc. (ARI), fluidized-bed catalytic oxidation system or approved equal was specified. The system configuration is shown in Figure 1.

Since operations commenced, the system has operated a total of 7,760 hours out of a possible 8,837 hours. This is equal to an operating efficiency of 87.8%. The system's downtime was attributed to maintenance, testing, system modifications, repairs, and experiments for a total of 1,077 hours. It should be noted that maintenance of the system attributed to only 3.8% of the total downtime, equaling approximately 5.8 h/wk.

Lombard et al. 85

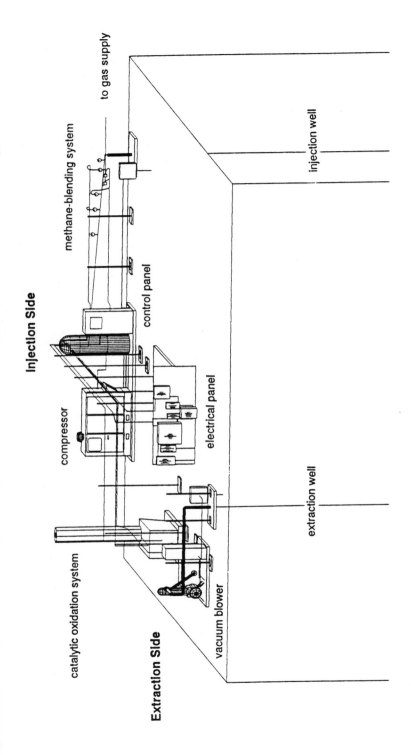

FIGURE 1. Isometric configuration of surface-mounted equipment for the in situ methanotrophic bioremediation demonstration.

SYSTEM COMPONENT DESCRIPTION

Methane-Blending System

The heart of the surface-mounted equipment supporting the methano-trophic bioremediation process is the methane-blending system (Figure 2). The system is designed to inject and blend a controlled flow of methane (natural gas) with the airstream, which is introduced into the lower horizontal injection well. The system's principal components are comprised of a Foxboro controller, Horiba gas analyzer, Foxboro/Jordan rotary actuator, and flow control valve, as described below.

Methane Gas Analyzer. The analyzer used for this project was an Horiba model PIR-2000 general purpose gas analyzer. It is a "precision gas analyzer based on nondispersive infrared ray absorption for contin-uously determining the concentration of a given component in a gaseous stream" (Horiba literature). It operates on an airstream of approximately 2 scfh (0.9 L/min) at a pressure less than 1 psig. For purposes of this

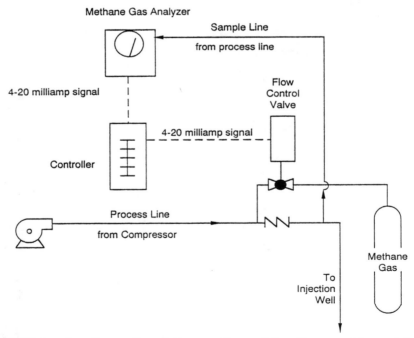

FIGURE 2. A schematic of the methane-blending system which is designed to inject and blend a controlled flow of methane (natural gas) with the airstream.

project, a small stream of gas was diverted from the process injection line to operate the gas analyzer.

The instrument initially was factory-calibrated for handling concentrations of methane between 0 and 100% of the LEL. The signal output from the analyzer is 4 to 20-milliamps, representing full span of the LEL. This signal is transmitted to the Foxboro controller.

The quality and performance of the Horiba Analyzer was very good, consistently producing precise and repeatable data. However, analyzers require factory recalibration on an annual basis and field calibration on a daily basis. The field calibration procedure requires two standard gases (zero and full span).

Foxboro Controller. The controller used for this project was a Foxboro 761C single station micro plus controller, which has a reputation for being a highly reliable controller. The Foxboro controller values for the proportional band, integral and derivative (PID) actions function by using a continuous monitoring adjustment process that measures the difference between an incoming signal and set point control variables (Foxboro literature). The real-time process dynamics vary the dampening and overshoot response effort based on controller-calculated parameters.

The controller reads the input signal from the analyzer and compares the indicated methane concentration with the set point concentration. The controller adjusts the concentration via the flow control valve to the indicated set point. This is done by sending a second 4 to 20-milliamp signal to the Foxboro/Jordan rotary actuator which positions the flow control valve to the required opening. If the indicated concentration is above or below the controllers' set point, the signal to the flow control valve is adjusted accordingly.

Foxboro/Jordan Rotary Actuator and Flow Control Valve. The Foxboro/Jordan rotary actuator receives the signal from the controller and mechanically adjusts the flow control valve. The valve, a segment style 1/4 turn ball valve, was selected because it has a fairly linear performance curve over the flow range anticipated.

All of the equipment and instrumentation devices in this control loop have operated well within the design parameters. Concentrations are controllable to a set point with a tolerance of ± 2% of the LEL (± 0.11% by volume). Nevertheless, although the system responded nicely to moderate changes, quick or sudden changes could push the system out of tolerance and trigger alarm conditions.

Oxidation of Contaminant Vapors

Treatment Options. Despite advances in environmental remediation technologies, there are still only two viable options available for treatment of off-gases from vapor extraction systems: (1) thermal type systems, comprised of catalytic oxidation or thermal catalytic oxidation; and (2) carbon absorption, which requires further treatment to destroy extracted contaminants. The larger share of this technology development has revolved around the improvement of catalytic materials, which now can handle a much greater variety of reaction-inhibiting compounds (Hardison & Dowd 1977).

These contaminant destruction methods are energy intensive and very expensive under the design requirements of today's vapor extraction projects (large volumes of air and concentrations of contaminants in the range of 0.5 to 1000 ppm). It could be argued that the energy consumed, associated resources used, and potential contamination produced may offset the environmental gains made by a vapor extraction project. Nevertheless, in most applications, the regulators (the U.S. Environmental Protection Agency [EPA] and the SCDHEC) require the contaminants to be removed from the airstream as directed by the Clean Air Act, 1991.

Thermal Catalytic Oxidation System Design

Site resources and design considerations for the vapor extraction dictated the use of a thermal catalytic oxidation system. Once the decision to use thermal catalytic oxidation had been made, several options required design engineering evaluation. Those included the power source (electric vs. propane/natural gas), air preheating, and the type of catalyst to use.

Power Source. Two options are available for heating: electricity and gas power. Given the choice, most engineers will agree that a gas-heated system is preferable from a safety and economic perspective. Oxidation systems require huge amounts of energy, as much as 200,000 BTUs per hour. For an electric system to produce these requirements, it is not unusual to operate a 480-V power system in excess of 250 amps. However, in some situations an electric system is the only option available.

Industrial settings, such as refineries, where the presence of an igniting flame is prohibited, rule out a gas-powered system. Electric systems, for these situations, can be made explosion proof and suitable to a Class I, Division 2 electrical areas. In some areas gas simply is not available, as was the case at SRS. Electric systems typically cost more. A more complex control system and a smaller market for these systems account for much of the price difference. Additionally, many of the manufacturers that make electric systems make them as a modification to a gas system, which further drives up the cost.

It is worth remembering that some of the power requirements can be reduced by the contaminant itself. BTU content of the contaminants can be used to provide some of the energy requirements. For this to be effective, the contaminant concentrations must be very large, certainly larger than the 200 to 400 ppm available on this project. Nevertheless, one should be careful in designing around this heat source, as concentrations usually fall off precipitously in the first weeks of extraction and continue to fall over the course of a successful project.

Air Preheating. Air preheating involves the use of a heat exchanger to heat the incoming extracted air using recovered hot exhaust gases. A heat exchanger can recover 40 to 50% of the waste heat, and will most certainly pay for itself in reduced energy costs. However, the increase in capital costs is significant and it should be carefully considered.

On projects where chlorinated contaminants are being destroyed there will be hydrochloric acid produced. The presence of the acid requires the use of materials that are more acid tolerant. These materials generally are very expensive and make the use of preheaters a more difficult decision to make. The duration of the project and the cost of energy will dictate the answer to this question. On the SRS project, preheating was not used. This decision was based in part on the short duration of the project (12 months) and the need to have the equipment fabricated and deployed quickly in the field.

Catalyst Selection. Treatment of contaminants containing halogens has been a dubious proposition at best, in recent years. Several elements can poison or inhibit the activity of the catalytic material. For example, halogens can quell the catalytic activity, but the effect can be reversed by removing the catalyst from the system and washing it to remove the contaminants. Nevertheless, when a halogen is absorbed to the catalyst surface it renders the material ineffective for the oxidation process. Halogen tolerance should be of primary concern in the selection of a catalytic material.

A second consideration for catalyst selection is whether to use a fixed catalyst bed or a fluidized bed. In a fixed catalyst bed, the material is attached to the surface of a substrate which has a high porosity. The contaminated air then passes through the substrate and comes in contact with the catalyst and reacts. In a fluidized bed, the catalyst is attached to small particles. A chamber is filled with particles and the contaminated air is passed through them. When a certain air velocity is attained, the particles become fluidized and begin to mix vigorously. This allows good contact between the contaminants and the catalyst material. One benefit of this method is that the catalyst material can easily be removed or added to the system for cleaning and replenishing purposes. Because all catalysts eventually loose their activity, this is a benefit for projects with long durations.

Several catalysts produced in recent years claim to be halogen tolerant, but their performance has not been firmly established. Toward that end, a second-generation, halogen-tolerant, fixed catalyst manufactured by Allied Signal Inc. was installed in the thermal oxidizer for testing its effectiveness in the destruction of chlorinated solvents. With the addition of the new catalyst, the operating temperature of the oxidizer was lowered to 825°F (441°C), resulting in an energy savings of more than 41%, while still maintaining a destructive efficiency of greater than 95%. A complete report of the results of this modification will be published at the completion of the project.

Vacuum-Blower and Compressor

The vacuum-blower and the compressor are the workhorses of the remediation system, operating continuously 24 hours a day. Both units turned out to be slightly overdesigned for the project and could be replaced with smaller units on a second generation system.

Vacuum-Blower. The vacuum-blower is a rotary positive-displacement Roots, Model URAI-56, and is powered by a direct-drive 20-HP electric motor. The vacuum-blower can pull a vacuum up to 10 in. (25.4 cm) Hg while producing a volume of 300 scfm (142 L/sec) of air. It is controlled by a Reliance Electric GP-2000 A-C VS controller, which allows for variable speed control of the 480-V motor. With this controller, the volume of air and the vacuum drawn on the extraction system can be regulated to meet the needs of the demonstration. For example, additional alternative off-gas treatment technologies can be tested simultaneously without reducing the primary project needs. The total cost of the vacuum-blower and control system is approximately $16,000.00. Although this is very expensive, the equipment has worked flawlessly since the project started.

Compressor. The compressor is a Quincy rotary screw assembly Model QNW-360-1, with an air-cooled oil cooler and a 75-HP, 480-V motor. The unit can supply 361 scfm (170 L/sec) of compressed air at 110 psig (50 kg/cm^2). The airstream goes to a 240-gal (908-L) air receiver tank that provides a reservoir of compressed air for the system to draw from, then through a regulator where the pressure and flowrate are adjusted to the injection well conditions. The injection flowrate of 200 scfm (94 L/sec) is manually controlled to ± 2%.

The injection well back pressure fluctuates between 20 to 30 psig (9 to 14 kg/cm^2), depending primarily on the soil moisture conditions at any particular time. Rainfall events are the major variable affecting the well back pressure. Other applications and soil types may have different back pressures, but this amount has proven to be in the range of typical values. This is unfortunate, for it is at a pressure range that borders on the lower efficiency range of the compressor. Compressors become less efficient and more costly to run at pressures and volumes such as this unit is experiencing. However, the compressor, like the blower, has performed consistently over the course of the project.

DISCUSSION

Instrumentation

For some engineers, there can never be enough instrumentation, and to some extent there may be strength to this argument. Inexpensive local reading gauges are readily available and can be incorporated into the design with a minimum of effort.

The thermal catalytic oxidation system used on the project was modified to correct instrumentation inadequacies. It was noted during start-up and testing that temperature gauges were not present at critical monitoring points: for example, areas where electrical power cables entered hot compartments, power feed attachments to heating electrodes, and the process airstream in front and behind the catalytic material.

By adding instrumentation to critical detection points, system deficiencies were detected and corrected prior to full-scale operation. Other problems associated with the system have been in the controls and power supply electronics. As mentioned earlier, natural gas was not available at the site, and an electric unit was required.

Electric units have a fairly complex control system which adjusts the 480-V, 3-phase current to keep the temperature within the specified limits. The Omron HL 2000 temperature controller, Model No. F5AX-A, was able to keep the temperature within 1 degree of the required set point. The temperature signal to the Omron controller, was from a type K thermocouple inserted in the combustion chamber of the oxidizer. A 4- to 20-milliamp control signal through a silicon-controlled rectifier (SCR) was used to power the silicon carbide resistance heaters. The heater unit also was equipped with a multiple-tap transformer control center to reduce the 480, 3-phase voltage as the heater elements aged. This action helps extend the life of the unit.

The area where the methane is blended is a Class I, Division 2 electrical zone, and equipment in that area must be intrinsically safe or explosion proof. To avoid having to supply equipment to meet those rigid specifications, the gas analyzer(s), controllers, and monitoring devices were placed outside of the area in a common enclosure (panel). The consolidation of the instrumentation with the high-voltage equipment proved to be an engineering challenge.

Mass flow of air and methane were attained by using two calibrated Kurz in-line mass flowmeters, Model Nos. 505 and 455. The explosion-proof flow sensors were positioned for in-line detection, transmitting flow data to the data logger and digital display located in the common control panel. When properly calibrated, the flowmeters gave extremely accurate readings. The readings were verified by gas chromatography.

Materials and Spare Parts

Construction codes and standards for systems carrying potentially hazardous materials are significantly more restrictive than for ambient air systems. Hydrochloric acid is created by the oxidation of chlorinated solvents. Although the site is well within the limits imposed by the air permit, acid is created and the materials used must be corrosion resistant in order to survive the project. Another consideration is to use materials that will not generate an additional wastestream by becoming a hazardous waste when the project is complete.

Systems for scrubbing out the acid and neutralizing it are available but can be costly. Additional permitting may be required, and possibilities exist for additional wastes to be generated. During the course of this project, several components were taken out of service or failed, for various reasons. A contingency of spare parts and key components is essential for a successful project.

Equipment Configuration

The main emphasis was placed on the relationship between the methane-blending system (Figure 2), the gas delivery system, and the gas analyzer, monitoring devices, and the electrical supply to the thermal catalytic oxidizer. In order to get the blended air/methane sample to the gas analyzer, a 20 ft (6 m) length of tubing had to be run from the blender mixing point to the common control panel. This put a lag of about 15 seconds between the time in which the gas concentration was adjusted by the flow control valve and the altered gas mixture reached the analyzer.

This delay forced the controller to use a rather slow-response algorithm for adjusting the loop to fluctuations in the gas concentrations. For that reason, it would have been desirable to shorten the length of the sample line, or in the absence of that possibility, use a larger diameter sample line, increasing the flowrate to the analyzer. During cold weather operations this delay in signal response resulted in frequent alarm conditions. Significant efforts to fine-tune the system were required.

Another relationship, that presented interface problems, was the housing of low-voltage, heat sensitive instrumentation and high-voltage, heat generating power supplies in the same control panel. Calibration or adjustment of the monitoring equipment required working in very close proximity to live 480-V power. In fact, the safe operating range (120°F [49°C]) of some devices, was pushed to its limit during hot weather. To eliminate potential safety hazards and component failures, separate control panels should be specified for low- and high-voltage components.

Weather Conditions

The Savannah River Site, located in South Carolina, gets frequent and sometimes severe thunderstorms during the summer months with numerous lightning strikes. In areas such as this, lightning protection is a cheap form of insurance and is easy to install. There were several failures in the power control system and instruments, many due to lightning. In fact, lightning strikes contributed to over 80 hours of the system's downtime. It is relatively inexpensive to obtain the services of a local electrical engineer, familiar with the area, to design lightning protection into the system controls and obtain the necessary site-specific equipment. Designing lightning protection by surveying the literature and trying to make an educated purchasing decision may give the designing engineer a false sense of security. It is best to consult with local professionals in this case and remember, the small amount of effort just described will, and has on this project, provided sufficient protection for most situations.

Heat and high humidity placed unnecessary stress on the sensitive equipment in the control panel. This was alleviated by installing an air-conditioning unit in the panel. Although simple, the field modification required additional power revisions. Cold weather also provided some challenging conditions. As stated, the SRS does not have a direct supply line for natural gas. To support the project with methane gas, two compressed natural gas (CNG) tube trailers were manufactured. Each trailer has a capacity of approximately 12,500 scf (354 m^3), at 3,000 psig (1,361 kg/cm^2).

To meet the injection pressure at the wellhead, the trailers were equipped with a multistage regulator system, capable of reducing the pressure from 3,000 to 40 psig (18 kg/cm^2). A general rule of thumb for CNG is, for every 100 psig (45 kg/cm^2) reduction in pressure, the temperature of the flowing gas is reduced by 70°F (21°C). The trailers came equipped with an ambient loop between the second and third or final-stage regulator. The loop is designed to prevent freezing in the final-stage regulator. The loop worked well during warm weather, but when outside air temperatures dropped below 40°F (4.5°C), the system failed due to frozen regulators and supply lines. An erratic gas flow resulted, triggering high methane concentrations and system shutdown. As of this date, this condition has not yet been resolved. Several proposals are under consideration. It would be a great advantage if a low-pressure CNG source were available at the site.

Safety Controls and Interlocks

The safe operation of the system must be paramount in the mind of the designer. Significant redundancies and safety interlocks must be designed into the system to prevent any unforeseen hazardous conditions from existing undetected. The SRS system includes audio and visual alarms, safety interlocks between major components (i.e., the methane injection system cannot function if the extraction system is not operational). There are overtemperature controls to prevent excessive heat buildup in the oxidizer and provide for unit shutdown in the event of a thermocouple burnout and mechanically operated solenoid valves that are normally closed if a power failure occurs.

Additional safety devices include off-gas low/high-pressure alarm switches, methane gas detection that shuts down the system if concentrations greater than the LEL are detected, and overcurrent protection with alarm and system shutdown capabilities. In the event of any malfunction, an emergency response telephone autodialing system, tied to the pager network of management, engineering, and operations staff, is activated.

SUMMARY

The events and conditions described in this paper are important to a successful project. Remember how important even the smallest missed detail can be in today's competitive economic climate. Small errors or inadvertent deletions can be very problematic and sometimes extremely costly.

The project was, without a doubt, a success from an engineering standpoint. The flexibility of the design to accommodate field modifications and the quick response to initial component and system failures made the endeavor successful. The total system downtime from field modifications and repairs, including the catalytic oxidizer modification, was only 214 hours of the total possible operating time. This represents only 19.8% of the total downtime. Significant knowledge has been gained about the reliability, compatibility, and performance of the system components in a unique environmental remediation application.

Elements of the system have been stressed to their limits, some to the point of failure. This action, although not intentional, provided data to design and build superior second-generation systems. The lessons learned from the research and development of this innovative technology should help to develop more efficient, cost-effective, and environmentally sound remediation programs. Future exploitation of these successfully tested technologies will provide information applicable to more difficult remediation problems where no proven methods exist.

ACKNOWLEDGMENTS

The information contained in this article was developed under the course of work under contract No.DE-AC09-89SR180035 with the U.S. Department of Energy. In addition, portions of this information were developed by Bechtel Savannah River Inc. and ECOVA Corporation as part of a subcontract with Westinghouse Savannah River Company.

REFERENCES

Eddy, C. A., B. B. Looney, J. M. Dougherty, T. C. Hazen, and D. S. Kaback. 1991. *Characterization of the Geology, Geochemistry, Hydrology and Microbiology of the In-Situ Air Stripping Demonstration Site at the Savannah River Site.* WSRC-RD-91-21. Westinghouse Savannah River Company, Savannah River Site, Aiken, SC.

Foxboro Company. 1985. *Exact Tuning with 760 Series Controllers,* Technical Information Publication, Bulletin No. TI 039-200.

Hardison, L. C., and E. J. Dowd. 1977. "Air Pollution Control: Emission Via Fluidized Bed Oxidation." *Chemical Engineering Progress,* Air Resources, Inc.

Hazen, T. C. 1991. *Test Plan for In Situ Bioremediation Demonstration of the Savannah River Integrated Demonstration Project.* DOE/OTD TTP No. SR

0566-01 (U). Westinghouse Savannah River Company, Savannah River Site, Aiken, SC.

Horiba Instruments Inc. 1987. Instrument Manual for Horiba Model PIR-2000. "General Purpose Infrared Gas Analyzer." Horiba Manual No. 090652, 4-87A.

Looney, B. B., T. C. Hazen, D. S. Kaback, and C. A. Eddy. 1991. *Full Scale Field Test of the In-Situ Air Stripping Process at the Savannah River Integrated Demonstration Test Site.* WSRC-91-22. Westinghouse Savannah River Company, Aiken, SC.

Looney B. B., and J. Rossabi. 1992. *Assessing DNAPL Contamination A/M Area, SRS: Phase I Results (U).* WSRC-RP-92-1302. Westinghouse Savannah River Company, Savannah River Site, Aiken, SC.

Marine, I. W., and H. W. Bledsoe. 1984. *M-Area Groundwater Investigation (Supplemental Data Summary).* Savannah River Laboratory Report DPSTD-84-112, Savannah River Plant, Aiken, SC.

BIOREMEDIATION BY GROUNDWATER CIRCULATION USING THE VACUUM-VAPORIZER-WELL (UVB) TECHNOLOGY: BASICS AND CASE STUDY

W. Buermann and G. Bott-Breuning

INTRODUCTION

Not only in the industrialized countries, but worldwide, the number of known groundwater and soil air contaminations by hydrocarbons; benzene, toluene, ethylbenzene, and xylenes (BTEX); pesticides; nitrates; etc., increases. Efficient, low-cost remediation techniques are needed.

A new method for the in situ remediation of groundwater and soil air is the vacuum-vaporizer-well (UVB) technology (German: Unterdruck-Verdampfer-Brunnen [UVB]; invented by B. Bernhardt; patents: IEG mbH, D-7410 Reutlingen). The disadvantages of groundwater remediation applying current pumping methods (groundwater lowering, limited yield, insufficient remediation) may be avoided if pumping and recharge take place in the same well. The UVB technology applies this circulation well concept.

The basics of hydromechanical theory are outlined in some detail (Buermann 1990, Bürmann 1991). Results of the field measurements conducted in Karlsruhe, Germany, to verify the UVB technology have been published briefly (Bürmann 1992, Bürmann & Wagner 1992) and are presented.

A case study on the bioremediation of pesticide (triazines)-contaminated groundwater is presented. Activated carbon is placed within the UVB well as a biofilter. A decrease in triazine concentrations in the groundwater is documented. An increase in the number of bacteria in the aquifer was observed and suggests a stimulation of biological processes. Development of metabolites within the activated carbon filter provides evidence of triazine biotransformation.

1-56670-084-1/94/$0.00 + $.50

Operation of the Vacuum-Vaporizer-Well: UVB Technology. The UVB produces a circulation flow within the surrounding groundwater, directed from the upper to the lower screening, as seen in Figure 1. Water is sucked into the lower screening, transported upwards inside the UVB by the water pump (air lift pump), and cleaned by fresh air in the stripping zone under below-atmospheric pressure before flowing out of the UVB through the upper screening. This all takes place without the water leaving the aquifer. If necessary, the groundwater is cleaned on site and directed back to the well. Soil air from the unsaturated zone of the aquifer may be sucked into the UVB through the upper screening and thus also may be cleaned. The contaminants in the stripping air are adsorbed by activated carbon. To avoid precipitation, the stripping air loop is closed. Thus contaminants that are not adsorbed can be kept from escaping into the atmosphere (Herrling et al. 1992).

In resting groundwater, circulation creates a permanent flow and consequently cleans the soil within the zone of the well, as all the circulating water flows through the well. Natural groundwater flow, which exists in most cases, deforms the circulation flow so that a portion of the water flowing toward the intake zone of the well may pass the well several times, due to the continual circulation flow, whereas the remainder of the water flows through the well only once. Therefore, the cleaning equipment of the UVB must be dimensioned so that one flow through the well is sufficient to ensure decontamination of the water.

Groundwater Flow around the UVB. The circulation flow depends on the natural groundwater flow, the water flowrate through the well, the water-saturated thickness of the aquifer (corresponding to the length of the well), the lengths of the lower and upper screenings, the outer radius of the well, and the horizontal and vertical conductivities of the aquifer (Buermann 1990).

The circulation flow may be influenced only by the design of the well itself, and in particular by the water flowrate. If existing wells must be used, water flowrate is the only means of control of the circulation flow.

In resting groundwater, the investigations give a theoretically unlimited zone of effect of the well. For a realistic judgment of the zone of effect, a radius around the well is chosen that contains a specific percentage of the total quantity of water flowing inside the well. The influence of the screening length is small. For realistic values of the anisotropy of the aquifer, the radius of effect is approximately 1.5 to 2 times the water-saturated aquifer thickness.

The circulation flow in moving groundwater shows two separating streamlines, at the bottom and at the top of the aquifer, similar to the

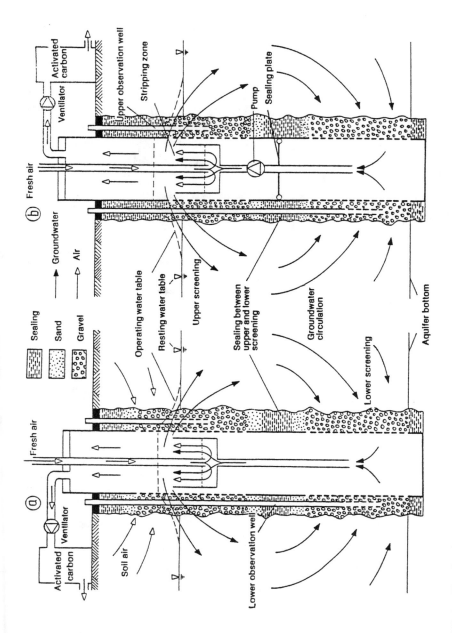

FIGURE 1.   Typical vacuum-vaporizer-well (UVB).

perfect well (Figure 2). In a well with upward flow, the lower separating streamline corresponds to the withdrawal well and the upper one to the infiltration well. Between these two separating streamlines at the lower and upper boundaries of the aquifer lies the separating stream surface of the flow around the well in the natural groundwater. This surface consists of spatial streamlines and shows a different contour in each horizontal section.

The dimension of the separating stream surface is characterized by the distance of the stagnation point S from the well. Figure 3 shows the water flowrate over the stagnation point distance of the upper separating streamline. The lower stagnation point distance gives the same curves for equal lengths in the lower and upper screening, and the curves remain essentially the same even for very different screening lengths. The smaller the ratio of vertical and horizontal conductivity, the greater the stagnation point distance and the influence zone of the well.

The water flowrate through the well rises more than proportional with the stagnation point distance. Therefore, instead of one single well of a large water flowrate, several wells of small rates may be useful.

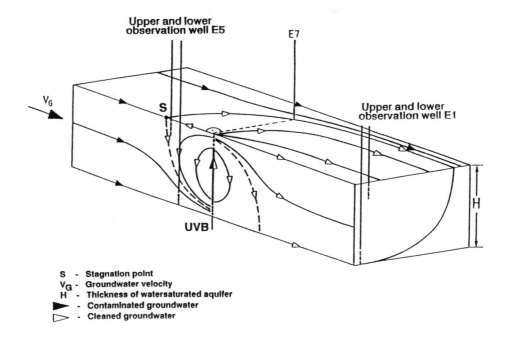

S   - Stagnation point
V_G - Groundwater velocity
H   - Thickness of watersaturated aquifer
▶ - Contaminated groundwater
▷ - Cleaned groundwater

FIGURE 2. Typical flow pattern of the vacuum-vaporizer-well (UVB) in natural groundwater flow.

CASE STUDY OF A BIOLOGICAL REMEDIATION

The UVB technology offers not only an innovative method of physically remediating contaminated sites, but also makes in situ biological remediation of groundwater possible. As a case study, a combined physical and biological remediation of groundwater containing pesticides (triazines) is presented (Figure 4).

The darcy velocity of the natural groundwater flow of 0.17 m/d, the water-saturated thickness of the aquifer of 6.6 m, the anisotropy k_V/k_H of 0.1, the screening length of 2 m, and the water flowrate inside the UVB of 4 m^3/h give the stagnation point distance of about 13 m in Figure 3.

Principle of Bioremediation. The principle behind every bioremediation is optimizing the environmental conditions for the naturally existing, already adapted microorganisms. Oxygen often is a limiting factor for aerobic degradation. The part of the aquifer where the UVB creates a continuous circular flow is regarded as an in situ bioreactor and is constantly supplied with oxygen-enriched water. Additional nutrients needed by the bacteria can easily be injected into the circulation flow that

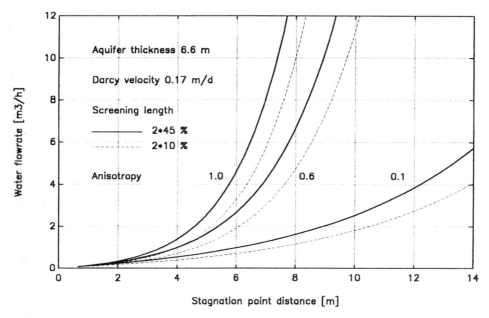

FIGURE 3. Water flowrate over stagnation point distance of the vacuum-vaporizer-well (UVB) in natural groundwater flow.

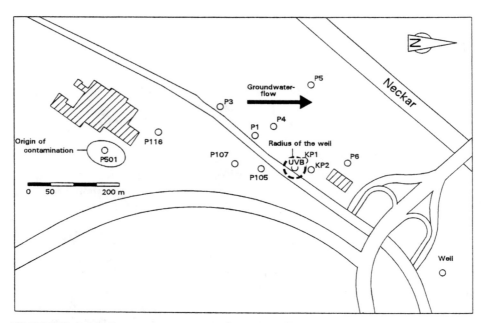

FIGURE 4. Schematic map of the contaminated site.

the UVB creates within the aquifer. These nutrients enable optimal conditions to be created for the microorganisms bound on grain surfaces.

In the case study presented in this paper, activated carbon was used as a biofilter within the UVB. The two variations shown in Figure 5 were tested. In both cases the contaminants and the triazine-degrading bacteria are adsorbed onto the activated carbon by constant circulation of contaminated groundwater in the well. This accumulation is a special advantage in cases with low contaminant concentrations or few bacteria in the groundwater. Adding specific nutrient supply for the bacteria to the biofilter is possible.

Results of the Triazine Remediation. In Figure 6, the concentration curves of the total triazines (atrazine, propazine, simazine, and triazine metabolites) entering and leaving the biofilter are depicted. The amount of triazines in the groundwater entering the activated carbon is higher than that leaving the biofilter. This decontamination is the result of adsorption of triazines onto and biological degradation processes within the activated carbon.

During biodegradation of triazines, various intermediates are formed (Cook 1987). These were detected in the aquifer before remediation with the UVB technique began. Figure 7 shows the concentration curve of one

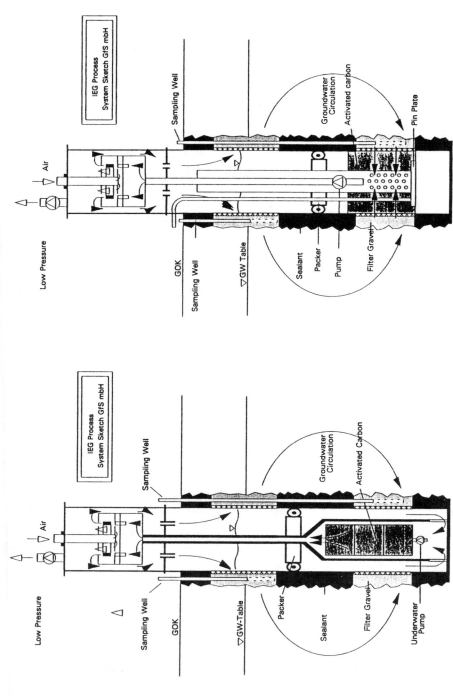

FIGURE 5. Version 1 (left) and version 2 (right) with the biofilter implemented (schematic).

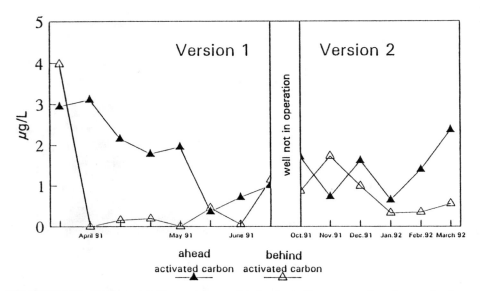

FIGURE 6. Concentration curve of triazines in groundwater entering and leaving the biofilter.

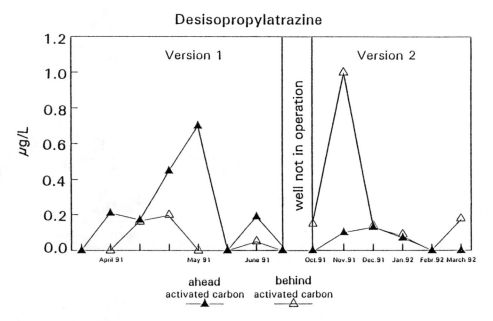

FIGURE 7. Metabolite concentration (desisopropylatrazine) in groundwater entering and leaving the biofilter.

of these metabolites, desisopropylatrazine, in groundwater before and after treatment by the activated carbon. The higher metabolite concentration behind the activated carbon indicates that further biological transformation of triazines occurs in the biofilter. This intermediate is further reduced by biodegradation. Figure 8 depicts the decrease of triazine concentrations in groundwater of the monitoring well KP1.

In addition to using intermediates as an indication of biodegradation, it is possible to count the number of bacteria in a sample. This was carried out by the colony-forming-units (CFU) method, in which bacteria are cultivated under aerobic conditions on a defined standard nutrient supplier. Table 1 shows the development of the number of bacteria in samples taken from various wells. Within 3 months the number of bacteria in monitoring well KP1 increased by a factor of 1,000, and the triazine concentration decreased accordingly. A biofilm developed on the activated carbon from April to June 1991. It was analyzed qualitatively and quantitatively. The number of CFUs was 7.7×10^4/g activated carbon, which is an enrichment compared to the number of bacteria (470 CFU/mL groundwater) ahead of the activated carbon biofilter.

CONCLUSIONS

The combined physical and biological remediation of triazine-contaminated groundwater using the UVB technology shows good success

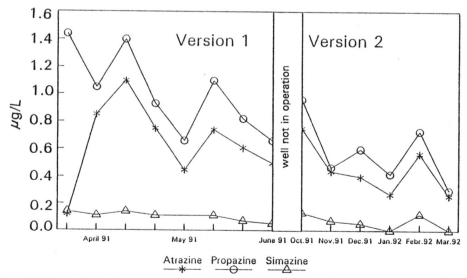

FIGURE 8. Triazine concentrations in the groundwater at monitoring well KP1.

TABLE 1. Development of bacteria (CFU/mL groundwater).

Date	Entering Activated Carbon	Leaving Activated Carbon	Monitoring Well KP1	Monitoring Well KP2
October 1991	$4.7*10^2$		$2.5*10^3$	
January 1992	$1.8*10^3$	$3.1*10^4$	$3.5*10^6$	$7.5*10^3$

in decreasing the triazine concentrations during remediation to date. The simultaneous increase in the number of bacteria in the aquifer suggests stimulation of biological processes. The development of metabolites and the increasing remediation rate within the activated carbon are evidence of biological triazine transformation. Further investigations include determination of degradation rate, looking for proof of specific triazine-degrading bacteria both in the aquifer and in the biofilter, and optimizing the biofilter.

ACKNOWLEDGMENTS

The authors gratefully acknowledge IEG mbH, D-7410 Reutlingen, for funding the investigations, and in particular, B. Bernhardt, IEG mbH, and many others for their work and their numerous helpful discussions and contributions concerning the UVB technology.

REFERENCES

Buermann, W. 1990. "Investigation of the circulation flow around a combined withdrawal and infiltration well for groundwater remediation demonstrated for the Underpressure-Vaporizer-Well (UVB)." In F. Arendt et al. (Eds.), *Contaminated Soil '90*, pp. 1045-1052. Third International KfK/TNO Conference on Contaminated Soil, Karlsruhe, December 10-14, 1990. Kluwer Academic Publ., Dordrecht, Boston, London.

Bürmann, W. 1991. "Zur Zirkulationsströmung am Unterdruck-Verdampfer-Brunnen (UVB)." In *Umweltplanung, Arbeits- und Umweltschutz 121*, pp. 64-80. Hessische Landesanstalt für Umweltschutz, Wiesbaden, Germany.

Bürmann, W. 1992. *Zusammenstellung der Meßergebnisse der Untersuchungen am Unterdruck-Verdampfer-Brunnen (UVB) im Karlsruher Versuchsfeld*. Bericht Nr. 698, Institut für Hydromechanik, Universität Karlsruhe, Karlsruhe, Germany.

Bürmann, W., and H. Wagner. 1992. "Remediation by groundwater and soil/air circulation in situ using the vacuum-vaporizer-well (UVB) technology." *Budapest '92 Proceedings*, International Symposium on Environmental Contamination in Central and Eastern Europe, October 12-16, Budapest, in press.

Cook, A. 1987. "Biodegradation of s-triazine xenobiotics." *FEMS Microb. Rev.* 46: 93-116.

Herrling, B., W. Buermann, and J. Stamm. 1992. "In-situ remediation of volatile contaminants in groundwater by a new systen of vacuum-vaporizer-wells (UVB)." In K. U. Weyer (Ed.), *Subsurface Contamination by Immiscible Fluids*, pp. 351-359. A. A. Balkema Publ., Rotterdam, Brookfield.

AIR SPARGING LABORATORY STUDY

M. A. Dahmani, D. P. Ahlfeld,
W. Ji, and M. Farrell

INTRODUCTION

Air sparging is a remediation technology which consists of injecting air below the groundwater table to clean soils and groundwater that are contaminated with volatile organic compounds. After injection into the water, the air migrates upward through the groundwater into the unsaturated zone. As the air passes through the contaminated water, volatile contaminants in the aqueous phase enter the airstream. The newly contaminated air is then extracted, typically using conventional vapor extraction technology. In effect, this system is an in situ "air stripper," i.e., the water is decontaminated in place by the mass transfer mechanism of volatilization. This approach has been used in a number of field applications in Europe and the USA (Bohler et al. 1990, Marley et al. 1992), but little work has been done on understanding the governing air-sparging process mechanisms.

The effectiveness and efficiency of this technology depend on several factors, primarily the detailed airflow pathways that will determine (1) the contaminated area that will be directly contacted by the air, and (2) the region that will be remediated indirectly through diffusion and other aqueous-phase transport processes. The project discussed here consisted of a laboratory study of the effect of grain size and soil heterogeneity on airflow through saturated porous media. Our observations of air travel through porous media in a sparging arrangement provide a first step in understanding airflow mechanisms and patterns. This knowledge is valuable in providing insight into the proper operation of a sparging system and is crucial for developing predictive mathematical models.

Due to the poor transparency of soils, glass beads have been used to improve flow visualization. Although glass beads, unlike soil particles, generally are smooth spheres, the mechanisms governing airflow through saturated porous media are expected to be similar at observation scales larger than the pore size.

1-56670-084-1/94/$0.00 + $.50

© 1994 by Lewis Publishers

EXPERIMENTAL PROCEDURE

Laboratory investigations of air bubbling or channeling through saturated porous media were conducted with a series of two-dimensional experiments. A Plexiglas™ tank, 88 cm high, 73 cm long, and 2.5 cm wide, was used. The thickness of the porous medium, equal to the width of the tank, rendered the experiments essentially two-dimensional to make flow visualization possible. This compression distorted the flow field by concentrating the air channels in a smaller volume, so quantitative measurements could not be extrapolated to the case of three-dimensional airflow in saturated porous media. Nevertheless, valuable qualitative information on the airflow patterns was obtained.

The porous media used were uniform, spherical, manufactured glass beads (specific gravity of 2.5 to 2.6) with sphere sizes of 4.0, 2.0, 0.75, 0.4, 0.3, and 0.2 mm diameter. The various mixtures of bead sizes represented a soil range of fine gravel to fine sand. The apparatus used for air injection consisted of a pressure regulator, a pressure gauge, a flowmeter, and a plenum. Air was injected at midpoint at the bottom of the tank. To enhance airflow visualization, a lighting panel was installed behind the tank. The increased light intensity drastically improved the contrast between water-saturated and air-saturated regions to enhance the quality of the photographs of the airflow patterns (Ji et al. 1993).

SUMMARY OF RESULTS

Each experiment was run twice. All the airflow patterns obtained were reproducible at the macroscale level. However, at the pore-scale level, the experiments were not reproducible due to the random packing of the glass beads. Preferential flow along the tank walls was minimized by tightly packing the beads.

Two distinct airflow regimes were observed. For homogeneous media (single size beads), the airflow regime type depends primarily upon grain size. For grain diameters of 4 mm and higher, corresponding to medium to coarse gravel (porosity = 35%), a bubbly flow regime exists in which individual air bubbles migrate through the water in response to buoyancy forces. For grain sizes of 0.75 mm or less, corresponding to sands, silts, and clays, the channeling regime maintains a dynamic equilibrium as long as the airflow remains constant. The transition between these two regimes appears to occur at ~2 mm grain size, within the range of airflow rates tested (up to 90 L/min). For either flow regime type, the airflow pattern was symmetric about the vertical axis and centered about the injection point. The radius of influence and channel density (number of channels per unit volume of soil) were clearly a function of air injection pressure.

The introduction of pore-scale and "mesoscale" heterogeneities [transition between microscale (pore-scale) heterogeneity and macroscale heterogeneity (layering)] yielded nonsymmetric airflow patterns. The bead mixture produced a medium that was homogeneous over the size of the tank (macroscale level) but heterogeneous within each small volume of the medium (pore-scale and mesoscale levels). The pore-scale heterogeneity is due to the different pore sizes produced from mixing various sizes of beads. The mesoscale heterogeneity is due to nonuniform mixing of various sizes of beads which produced fine stratification (< 2 mm in thickness) in some areas of the tank. The asymmetry of the airflow pattern apparently resulted from minor variations in permeabilities and capillary resistances induced by these heterogeneities. Therefore, a symmetric air distribution pattern may never occur in a natural subsurface soil due to the small-scale heterogeneities present even in soils that appear, at a large scale, to be homogeneous. Determining the influence region in such media becomes more difficult due to the asymmetry of the airflow pattern.

The larger scale heterogeneities, i.e., strata of differing permeabilities, produced airflow patterns strongly controlled by the contrasts in permeability (air flowed through higher permeability strata). Depending on layer thickness, layer areal extent, and injected air pressure, some air migrated through low-permeability lenses to the area above the lens. If this had not occurred, the portions of the porous medium directly above the low-permeability lenses would not have had direct contact with the injected air. Without direct contact, the efficiency of an air sparging system is significantly compromised. Contaminants in these isolated zones, whether in the aqueous, nonaqueous, or sorbed phase, would have to migrate to the air channel by convective-diffusive mechanisms before full remediation can occur. Thus, in remedial settings in which significant discontinuous stratification occurs, air sparging may have limited ability to efficiently reach contaminants present above and within low-permeability lenses.

The study also demonstrated the importance of air channel formation relative to contaminant removal efficiency. Volatile contaminants present near an air channel path should be readily removed within hours to days, but these experiments have shown that air channels cannot realistically traverse the entire contaminated region in a tree-like pattern because of soil heterogeneities. Thus, the volatile contaminants present in the bulk of the water phase can be removed only through diffusion to nearby air channels. The transport mechanism by which contaminants move from areas of the aquifer between air channels to the air/water interface then becomes the rate-limiting factor for air sparging. The experiments showed that the most homogeneous porous media produced the highest channel density within the air plume, indicating that media that are very homogeneous would be better candidates for air sparging.

In layered media, where dense, nonaqueous-phase liquids (DNAPLs) may be trapped on top of low-permeability strata, horizontal airflow beneath and vertical airflow around these strata will induce water flow around the DNAPL to increase the solubilization of contaminants into the water phase. The question remains whether these currents can induce sufficient contaminant transport to the air phase to significantly enhance the air sparging process efficiency.

REFERENCES

Bohler, U., J. Brauns, H. Hotzl, and M. Nahold. 1990. "Air Injection and Soil Air Extraction as a Combined Method for Cleaning Contaminated Sites. Observations From Test Sites in Sediments and Solid Rocks." In F. Arendt, M. Hinsenveld, and W. J. Van Den Brink (Eds.), *Contaminated Soils*, pp. 1039-1044.

Ji, W., M. A. Dahmani, D. Ahlfeld, J. D. Lin, and E. Hill. 1992. "Laboratory Study of Air Sparging: Air Flow Visualization." Submitted to *Groundwater Monitoring Review*.

Marley, M., D. J. Hazenbrouck, and M. T. Walsh. Spring 1992. "The Application of In-Situ Air Sparging as an Innovative Soils and Ground Water Remediation Technology." *Groundwater Monitoring Review*.

MICROBIAL AND CARBON DIOXIDE ASPECTS OF OPERATING AIR-SPARGING SITES

J. F. Billings, A. I. Cooley, and G. K. Billings

INTRODUCTION

Remediation of light, nonaqueous-phase liquids (LNAPLs) such as petroleum hydrocarbons, is a major problem throughout the world. The most promising methodology for cradle-to-grave cleanup of these contaminants appears to be enhanced bioremediation, wherein the contaminants are biologically decomposed, or mineralized into the by-products of carbon dioxide and water.

This paper provides general observations of microbial and carbon dioxide aspects at several air-sparging sites. Data are presented where a particular type of air-sparging was applied. The data suggest the predominant form of contaminant degradation at air-sparging sites is bioremediation. The biologic component is so prevalent, the authors prefer to think of a properly managed air-sparging system as a bio-sparging system.

REMEDIATION TECHNOLOGY

The air-sparging technology discussed in this paper, consists of a vapor extraction pump(s) connected to a series of extraction wells completed in the unsaturated zone. Air injection wells are completed beneath the water table and connected to an air-injection pump(s). The only aboveground feature is a Vapor Control Unit (VCU), which houses these pumps and ancillary equipment. Figure 1 presents a cross-sectional schematic of the SVVS™, the trade name of the air-sparging technology discussed in this paper.

Operation of the SVVS™ implements integrated remediation of both saturated and unsaturated soil and groundwater contamination (Ardito &

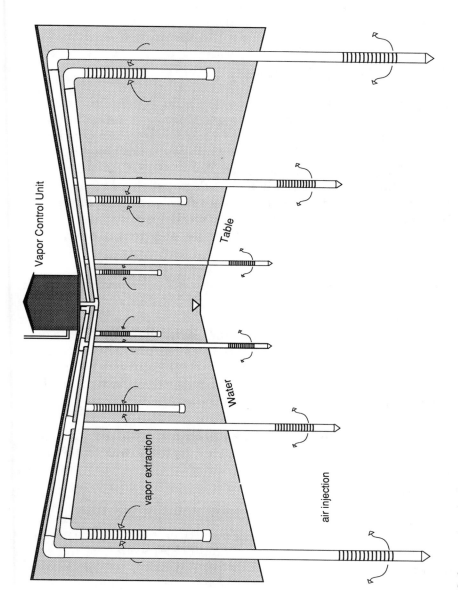

FIGURE 1. Schematic of SVVS™.

Billings 1990). Types of data collected during operation of an SVVS™ unit include hydrocarbon contaminants (vapor, water, soil), carbon dioxide (CO_2—vapor), dissolved oxygen (DO—water), oxygen (O_2—vapor), flow (vapor), and microbial populations (soil).

REMEDIATION RESULTS

The following case histories show probable cause-and-effect relationships of microbial and CO_2 aspects at air-sparging sites. Hence, increases in microbial populations with decreases in soil contamination, increases in DO with decreases in groundwater contamination, increases in hydrocarbon remediation with air-sparging compared to vapor extraction only, and determinations of the contribution of biodegradation are all presented. It should be remembered that none of these sites was a test funded for the purpose of researching the remediation mechanism, instead funding was for cleanup of the entire site. Consequently, only general operational trends are presented here.

Isleta Site. Air-sparging technology enhances biodegradation through the addition of DO into groundwater (Cooley & Billings 1992). The oxygen is used by the indigenous microbes in the consumption of the carbon source (hydrocarbon contamination) yielding the by-products of CO_2 and water. At the Isleta site, DO and dissolved-phase hydrocarbons were monitored during application of the SVVS™ biosparging technology. Figures 2 and 3 show increases in DO and decreases in hydrocarbon concentrations respectively, from two key monitoring wells, for 9 months of SVVS™ operation.

In all of the biosparging sites operated by the authors, detectable levels of volatile compounds (benzene, toluene, etc.) in groundwater have never been noted when DO is greater than 3.5 mg/L. Use of DO to monitor biosparging systems is useful in evaluating system performance.

Bass Site. A fundamental concept of biosparging is that by increasing microbial populations soil contamination will decrease. Microbe populations and soil total petroleum hydrocarbons (TPH) were determined at various locations throughout the Bass site prior to system implementation. Microbe populations averaged about 6×10^4 organisms/gm wt. of soil, and soil TPH concentrations averaged 2,300 mg/kg prior to system initiation. At periodic intervals during system operations the areas were resampled, measuring microbial populations and soil TPH.

Microbial populations increased about 7-fold during the monitoring period as shown on Figure 4. Figure 4 represents the average of the

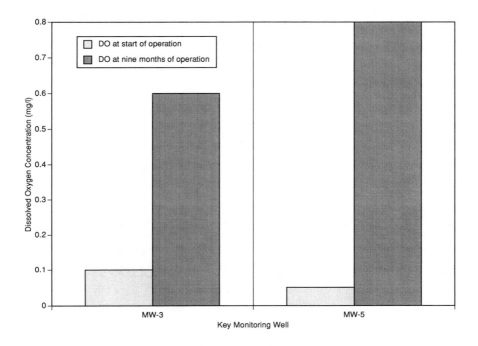

FIGURE 2. Isleta Site — increases in dissolved oxygen.

microbial populations taken from the 8 locations during the respective sampling period. Soil TPH concentrations decreased from an average of 2,300 mg/kg to about 800 mg/kg during the same periods (see Figure 5). Carbon dioxide concentrations in the vent gas averaged around 1 to 2% (volume) through the sampling periods. The ambient atmospheric concentration of CO_2 is about 0.03% (volume). It is therefore reasonable to conclude that, with oxygen being supplied to the subsurface, microbial populations increased metabolizing the hydrocarbons and concurrent decreases in soil TPH were noted.

Southwest Site. Following removal of an underground storage tank, an investigation at the Southwest site showed a large dissolved-phase plume. Groundwater at the site is relatively deep, on the order of 50 ft (15 m) below land surface. During the initial phases of SVVS™ operation, the system was operated as a vapor-extraction unit only (i.e., air injection was not initiated). Measurements were made of organic vapors and CO_2 in the vent gas. Injection of air was then initiated. Simultaneous air injection and vapor extraction was continued for some time. Finally, air injection was terminated leaving only vapor extraction operational.

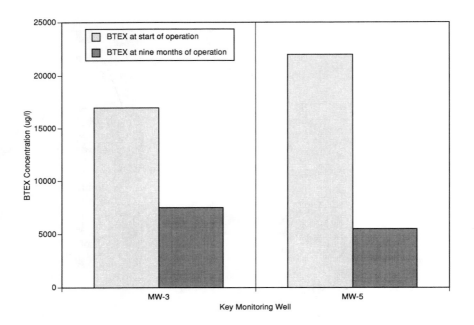

FIGURE 3. Isleta Site — decreases in dissolved-phase BTEX.

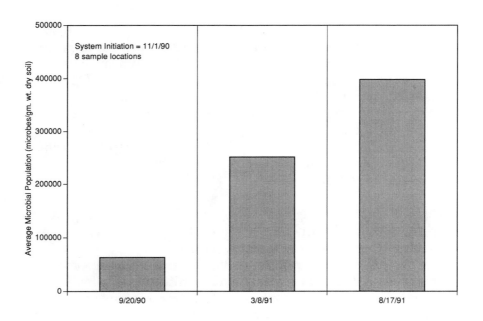

FIGURE 4. Isleta Site — increases in microbial populations.

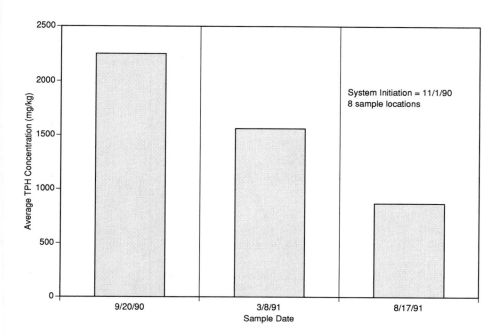

FIGURE 5. Bass Site — decreases in soil TPH.

Concentrations of volatile organic compounds (VOCs) and CO_2 versus time are shown on Figure 6. With initiation of air injection, a significant increase in remediation rate was noted as organic vapors nearly doubled. Similar results have been reported in the literature (Marley, 1991). The increase in the remediation rate is not solely attributable to volatilization, as CO_2 also increased with air injection from less than 1% (volume) to more than 6% (volume), close to an order of magnitude. Extraction flow remained the same during these phases, so the mass of contaminant remediation significantly increased with air injection.

Conservancy Site. An underground storage tank rupture released a known quantity of fuel into the shallow, water-table subsurface at the Conservancy site. Depth to groundwater at the site is about 9 ft (2.7 m), and the subsurface is characterized by fine-grained sands and clayey silts. Abatement measures included soil excavation around the tanks and free-phase product pumping. The remaining mass of contaminants was determined to be approximately 2,000 gallons (7,750 L), existing in soil residual, soil vapor, and dissolved-phase states. The SVVS™ was installed and initiated, and within 11 months of operation an estimated additional

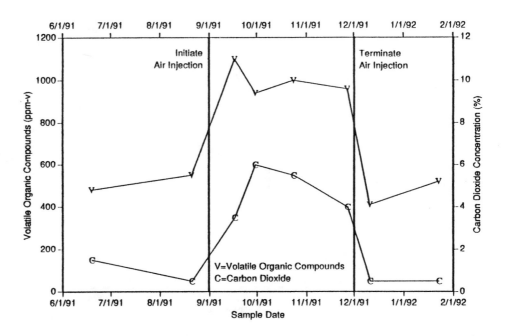

FIGURE 6. Southwest Site — CO₂ and VOC versus time.

1,100 gallons (4,164 L) of equivalent hydrocarbon contamination had been removed.

Hydrocarbon contamination removed by the SVVS™ during the first 11 months of operation was determined by two methods: (1) mass of hydrocarbons exiting the vent stack was computed as the sum of organic vapors and equivalent hydrocarbons converted from CO_2; and (2) using a mass-balance approach, the amount of hydrocarbons remaining in the subsurface was estimated from multiple soil borings and water-quality samples and then compared to that remaining after the mitigation steps (i.e., 2,000 gallons or, 7,750 L). Both methodologies suggested approximately 1,100 gallons (4,164 L) of equivalent hydrocarbon contamination had been removed by the SVVS™. Of the remediated quantity, about 770 gallons (2,914 L) or 70% was through the process of bioremediation (converted CO_2 measurements), with 330 gallons (1,249 L) removed through volatilization. This corresponds well with bioremediation to volatilization rates reported for bioventing sites (Miller & Hinchee 1990). As the system continues to operate, the bioremediation percentage will increase as few organic vapors currently are being extracted, but CO_2 in the vent gas is remaining stable at about 0.3%.

GENERAL OBSERVATIONS

- Dissolved oxygen concentrations increase when biosparging is activated, and concurrent decreases in dissolved-phase hydrocarbons are noted.
- Populations of indigenous, aerobic microbes increase several-fold when air, and consequently oxygen, is injected into the subsurface.
- Hydrocarbon concentrations of the soil decrease when microbial populations increase.
- Biosparging is considerably more effective in removing hydrocarbons than is vapor extraction alone, for sites with contaminated soils and groundwater.
- Biosparging increases the rate of biodegradation by providing a continuous source of oxygen directly into the source area.
- Biodegradation accounts for about 75% of the total remediation seen at biosparging sites.
- The term biosparging may be more appropriate than air-sparging, because the sites routinely remediate more hydrocarbons as CO_2 than as volatile compounds.

REFERENCES

Ardito, C. F., and J. F. Billings. 1990. "Alternative Remediation Strategies: The Subsurface Volatilization and Ventilation System." In *Proceedings of Petroleum Hydrocarbons and Organic Chemicals in Groundwater; Prevention, Detection, and Restoration.* Association of Groundwater Scientists and Engineers and American Petroleum Institute, pp. 281-296, Houston, TX.

Cooley, A. I., and G. K. Billings. 1992. "Integrated Technology for In Situ Groundwater Remediation." In *Air & Waste Management Association 85th Annual Meeting.* Kansas City, MO.

Marley, M. C. 1991. "Air Sparging in Conjunction with Vapor Extraction for Source Removal at VOC Spill Sites." In *Proceedings of Fifth National Outdoor Action Conference on Aquifer Restoration, Groundwater Monitoring and Geophysical Methods.* National Association of Groundwater Scientists.

Miller, R. N., R. E. Hinchee, C. M. Vogel, et al. 1990. "A Field Scale Investigation of Enhanced Petroleum Biodegradation in the Vadose Zone at Tyndall AFB, Florida." In *Proceedings of Petroleum Hydrocarbons and Organic Chemicals in Groundwater: Prevention, Detection, and Restoration.* Association of Groundwater Scientists and Engineers and American Petroleum Institute, pp. 339-349, Houston, TX.

GROUNDWATER CIRCULATION DUE TO AIR INJECTION: GENERAL ASPECTS AND EXPERIMENTAL RESULTS FOR LAYERED SUBSOIL

K. Wehrle and J. Brauns

INTRODUCTION

In recent years, groundwater contamination has become an issue of increasing importance. Conventional groundwater remediation includes pumping contaminated water out of the subsurface and treating on site by chemical, physical, or biological methods. For highly volatile organic compounds (e.g., CHC), air stripping and adsorption by activated carbon are common technologies.

There are also attempts to remediate in situ by transferring the contaminant from the water to the air within the subsurface (air sparging and related technologies). Bruckner (1987) presents some of the first applications of this technique, which he called "compressed air injection," and shows that the injection of air into the saturated zone in combination with soil vapor extraction (SVE) can cause significant mobilization of volatile groundwater contaminants. The injected air rises in the soil pores up to the unsaturated zone, thus transferring the contaminants from the dissolved or adsorbed phase to the vapor phase. Once in the unsaturated zone, the vapors can be captured by SVE.

Other in situ techniques that make use of groundwater circulation also have been established. A summarized description, including comments for users of groundwater circulation wells, is given by Zipfel and Kirschbauer (1992). The remediation well generally has two screens, one in the upper part of the well and the other at the bottom. By means of technical installations such as pumps, the water inside the well is moved from one screen to the other to create groundwater circulation. Cleaning the water involves additional installed equipment and is accomplished through other means (mostly by air stripping).

1-56670-084-1/94/$0.00 + $.50

The reliability of these in situ techniques depends to a great extent on both the subsurface and operating conditions. The following conditions must be met in order to achieve remediation:

1. The cleaning unit (e.g., stripping) must be efficient enough to ensure that the water flowing out of the well is clean.
2. Negative effects due to contact between the water and air, such as mobilization of heavy metals or precipitation and clogging, must be prevented.
3. The groundwater circulation zone must be wide enough to bring the contamination to the remediation well but prevent contaminants spreading away from the well.

In cases of air injection into the saturated zone combined with SVE:

4. Air entry into the unsaturated zone must be limited to the area influenced by the SVE. Spreading of contaminants must be prevented.

LABORATORY-SCALE EXPERIMENTS OF GROUNDWATER CIRCULATION

Previous Investigations. Previous laboratory-scale experiments have shown the influence of soil grading curves on the entry and movement of air in the soil (Brauns & Wehrle 1989). A distinction must be made between soils in which the pores allow the ascent of air by hydraulic uplift (uniform soils coarser than coarse sand) and soils in which the pores strongly oppose the entry of air. For fine soils, the injection pressure can cause air movement which is limited to distinct pathways where particle movement (suffusion) can take place. For coarse soils, the ascent is limited to a narrow area around the injection well. The air ascent in coarse soils induces groundwater movement in the form of circulation. Some results for homogeneous and for layered aquifers are given by Wehrle (1990).

The air injection results described so far are limited to aquifers that do not have layers preventing the ascent of air to the unsaturated zone. Air injection inside a well, as described in this paper, expands the applicability of this technique to aquifers that also contain fine-grained layers.

Materials and Methods. The experiments were carried out in a thin, box-shaped glass model (2.1 m long, 0.6 m deep, 0.1 m wide) so that groundwater flow was two-dimensional. The air injection well was made from a slotted pipe (inner diameter 30 mm), which was divided in two lengthwise and fixed to the front panel at the left end of the model. The well reached from the bottom to the top of the model. The injected air was run through a pipe (inner diameter 4 mm), which could be inserted into the well to any desired depth. At the front panel of the model, 20 color tracer input points were installed in 5 columns (I to V) and 4 lines (1 to 4). The color tracer indicated the flowlines. Documentation of the flowlines was carried out photographically. The velocities were determined by measuring the time the tracer needed to traverse a given distance. The aquifer consisted of a narrow graded gravel (2 to 3 mm). Modifications to the homogeneous case (case A) were carried out with an additional layer of sand (cases B1, B2, and B3), silt (case C), or silt with a window filled with gravel (case D) in the middle of the aquifer. The air injection to induce groundwater circulation was maintained at 80 L/h.

Results. In Figure 1, the flowlines of the air injection experiments for four different aquifer models are depicted. They start out from the 20 color tracer input points and are shown as far as they could be followed. During these experiments, the air injection pipe was placed 5 cm above the bottom.

In case A (homogeneous aquifer), the circulation flow reached the entire aquifer but the velocities (see Table 1) decreased rapidly with increasing distance from the well. All flowlines reached the well below the third line of tracer input points. Tracing the flowlines to their starting point shows that they all started out above the second line of tracer input points.

In case B1 with a 5-cm sand layer (0.1 to 1.0 mm), the groundwater flow passed through the sand layer. Every flowline spread further from the well than in case A, and the velocities at the tracer input points were higher except in the first column and at the top of the second column.

In case C with a 5-cm silt layer (<0.1 mm), the flowlines did not pass through the silt layer. There were two separate circulation paths, above and below the layer, but neither reached very far. The velocities at the tracer input points were higher in the bottom half than in the top half. The highest velocity (1.3 cm/s) measured during all experiments was recorded in this case in the first column below the silt layer. The flowlines starting out from the fourth and fifth columns could not be determined, because tracer movement was dominated by other effects such as density forces and/or temperature effects.

In case D, a 9-cm-wide window filled with fine gravel was opened in the silt layer 1m from the well. All flowlines passed through this window. Between the well and the window the flowlines were approximately horizontal, and considerable velocities were reached. Beyond the window the influence was slight.

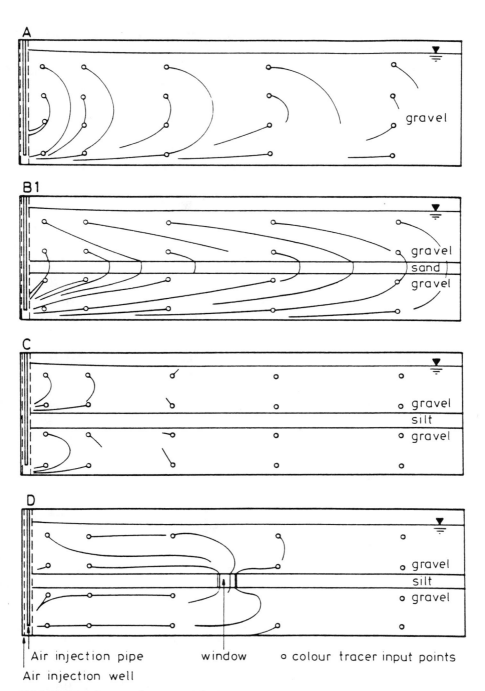

FIGURE 1. Groundwater circulation due to air injection inside a well. Flowlines for four different aquifer compositions.

**TABLE 1. Velocities (cm/h) of the groundwater circulation, measured
at the color tracer input points.**

	line\ column	I	II	III	IV	V
case A	1	1.0	0.42	0.04	n.a.	n.a.
	2	1.0	0.36	0.04	n.a.	n.a.
	3	1.0	0.25	0.04	n.a.	n.a.
	4	1.0	0.25	0.04	n.a.	n.a.
case B1	1	0.50	0.25	0.14	0.13	0.03
	2	0.83	0.63	0.25	0.13	0.03
	3	0.20	0.30	0.20	0.11	0.03
	4	0.33	0.50	0.17	0.10	0.03
case B2	1	n.a.	0.25	0.10	0.05	0.02
	2	0.28	0.24	0.16	0.10	0.02
	3	0.42	0.22	0.17	0.063	0.02
	4	0.50	0.18	0.17	0.067	0.02
case B3	1	n.a.	0.09	0.028	0.021	0.009
	2	0.08	0.10	0.033	0.023	0.009
	3	0.50	0.033	0.05	0.021	0.009
	4	0.50	0.09	0.042	0.021	0.008
case C	1	0.16	0.021	0.003	n.a.	n.a.
	2	0.20	0.017	0.002	n.a.	n.a.
	3	1.3	0.037	0.003	n.a.	n.a.
	4	1.3	0.059	0.003	n.a.	n.a.
case D	1	n.a.	0.33	0.42	0.028	n.a.
	2	1.0	0.33	0.33	0.014	n.a.
	3	0.83	0.46	0.25	0.028	n.a.
	4	1.0	0.50	0.25	0.021	n.a.

n.a. = not available

In Figure 2 two, variations of case B1 are depicted. The air injection
rate remained at 80 L/h, but the distance between the model bottom and
the air outlet of the injection pipe was varied (20 cm above the bottom
in case B2 and 35 cm in case B3). The results show that the flow decreased as the injection pipe was raised. The flowlines look very similar
in all cases. They varied somewhat near the well where it can be seen

that the area of inflow to the well increased as the air input pipe was raised and the outflow area shrank.

The influence of the air injection pipe position also was observed in cases A, C, and D (not shown). As in case B, the higher pipe positions caused smaller velocities and smaller outflow areas from the well. In case C, with air injection 20 cm above the bottom, the flowline starting at the point I,4 changed flow direction and reached the well just below the silt layer.

CONCLUSIONS

With these model experiments it is shown that air injection inside a fully screened well induces groundwater circulation. The air bubbles ascending inside the well cause a difference in potential between the well and the surrounding aquifer because of impulse flux forces and the different densities between the air-water mixture and the water. The value and sign of this difference determine whether inflow or outflow takes place at the well. In case A, with a homogeneous aquifer, the difference in potential

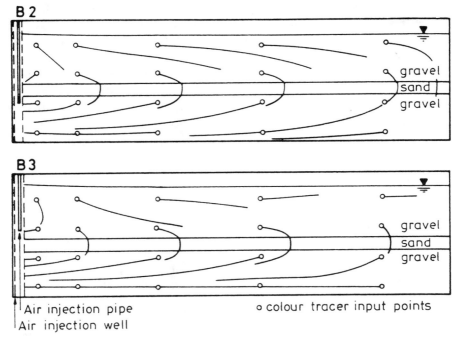

FIGURE 2. Groundwater circulation due to air injection inside a well. Flowlines for cases with different positions of the air injection pipe.

is a continuous function. Its greatest positive value (outflow from the well) is at the groundwater surface, and its greatest negative value (inflow into the well) is at the bottom. The gradient of the function grows with the distance from the bottom because of the decreasing density of the air-water mixture. A stagnation point occurs somewhere in between where the potential difference equals zero. For continuity reasons, the inflow rate equals the outflow rate.

The potential difference function shows another form in cases with a lower conductivity layer in the middle of the aquifer. Such a layer leads to a comparatively large potential "jump" between the top and the bottom of this layer where the size of the "jump" depends on the difference in conductivity between the layers. The function for the potential difference between the well and the surrounding aquifer also becomes discontinuous and can, in some cases, equal zero twice. Therefore two stagnation points can occur. The potential "jump" is responsible for the flow passing through the central layer and for the appearance of two separate circulations. The ratio of the conductivities of the different soils determines which effect becomes predominant. In case B, with a conductivity ratio of about $k_{gravel}/k_{sand}=10^2$, the flow through the central layer was predominant; in case C ($k_{gravel}/k_{silt}=10^6$), the appearance of separate circulations was predominant. In case D, the flow through the window was predominant but there were also signs of separate circulations in the first column of tracer input points. The flowline from the second line (I,2) goes directly to the well, and the flowlines from the third line (I,3; II,3) go downward near the well. It therefore seems that outflow from the well takes place just below the central layer.

The technique of air injection inside a well can be used in principle to create groundwater circulations in layered aquifers. The risk of spreading contaminants as a result of air injection is comparatively small, because layers with low conductivity cause separate circulations in layers with higher conductivity.

REFERENCES

Brauns, J., and K. Wehrle. 1989. "Untersuchung der Drucklufteinblasung in die gesättigte Bodenzone (In-Situ-Strippen), Modellversuche." Final report on behalf of the Landesanstalt für Umweltschutz (LfU), Karlsruhe, Germany (unpubl.).

Bruckner, F. 1987. "Sanierung von mit leichtflüchtigen chlorierten Kohlenwasserstoffen (LCKW) kontaminierten Grundwasserleitern mittels Kombination von Drucklufteinblasung und Bodenluftabsaugung." *Korrespondenz Abwasser* 34: 358-362.

Wehrle, K. 1990. "In-Situ Cleaning of Contaminated Sites: Model-Scale Experiments using the Air Injection (In-Situ Stripping) Method in Granular Soils." In F. Arendt et al. (Eds.), *Contaminated Soil '90*, pp. 1061-1062. Kluwer Academic Publishers, Boston.

Zipfel, K., and Kirschbauer, L. 1992. "Der Grundwasserzirkulationsbrunnen (GZB), Funktionsweise, bisherige Erfahrungen und Empfehlungen für den Anwender." Final report on behalf of the Landesanstalt für Umweltschutz (LfU), Karlsruhe, Germany (unpubl.).

APPLICATIONS OF IN SITU SOIL VAPOR EXTRACTION AND AIR INJECTION

C.G.J.M. Pijls, L.G.C.M. Urlings,
H.B.R.J. van Vree, and F. Spuij

INTRODUCTION

In the industrialized countries, the extent of soil contamination is enormous. The overall costs of soil remedial action in the Netherlands are estimated to be in excess of $30 billion. The costs of soil remedial action (excavation) and improving gasoline stations amount to approximately $750 million. Excavating the contaminated soil is a very effective method of removing the contamination.

The presence of infrastructure above and below the ground, which often occurs in sites situated in city centers and industrial estates is unfavorable to conventional excavation techniques. A powerful new technology for remediation which favors such sites is available with in situ techniques (Coffa et al. 1991, Urlings et al. 1991). This paper considers soil vapor extraction (SVE) and air injection (air sparging) as remedial techniques in general, and in particular for the air-based biodegradation of gasoline hydrocarbons.

REMEDIAL TECHNOLOGY

Soil Vapor Extraction (SVE). By creating negative pressure gradients in a series of zones within unsaturated soil, subsurface airflow is induced. This flow volatilizes the contaminants present in the unsaturated soil. This process, in theory, continues until all the volatile components have been removed. The extraction wells are individually connected to the transfer pipes, and are then manifolded into a vacuum unit.

The withdrawn soil vapor is often treated by activated carbon adsorption or catalytic incineration. The groundwater is usually treated by stripping and/or activated carbon adsorption. To minimize treatment costs of both the groundwater and the soil vapor, TAUW Infra Consult

B.V. has applied (since 1989) a biological system (BIOPUR®) for combined aerobic treatment of the groundwater and soil vapor. The system is patented in the USA and a patent is pending for Europe.

SVE, Air Injection, and Related New Technologies. Experience with soil vacuum extraction (SVE) is so vast that it can be considered a proven technology. The treatment costs of withdrawing soil vapor are substantial, usually more than 50% of the total remedial costs (Hinchee & Miller 1990). Air-based biodegradation is applied to reduce SVE costs by in situ biodegradation of the contaminants. This is a new innovative technique, as noted in an international review of in situ bioreclamation practice (Staps 1990). Only 2 of the 23 sites studied used air-based technology; most of the applied bioreclamations were water based. In recent literature (Miller et al. 1991, Hinchee & Miller 1990) air-enhanced biodegradation is described, mainly focused on the remediation of hydrocarbons.

Air injection (air sparging) is a relatively new technique for the remediation of volatile (chlorinated) hydrocarbons in the saturated zone. Only a few applications have been described in the literature. Air injection, as a means of delivering oxygen to the aquifer, to enhance bioremediation, is a new application of air injection. Using this technique, oxygen can spread quickly and more effectively in the saturated zone than by using extraction and infiltration wells.

Apart from the advantage of a lower hydrocarbon concentration in the withdrawn soil gas, significantly less carrier medium is needed if air is used as an oxygen carrier. Apart from promoting the biodegradation, air injection will also increase the stripping of volatile compounds from the groundwater.

An indication of the biological activity in the soil is given by several biological parameters such as Oxygen Uptake Rate (OUR) and CO_2 production ratio.

SITE 1: IN SITU SOIL VAPOR EXTRACTION

Characteristics. During soil remediation of a gasoline station, it became apparent that contaminants were present beneath a provincial road. Excavation of this part of the contaminated soil was not feasible due to financial and technical (traffic) reasons. The most favorable solution was to combine an SVE system with biostimulation. This system had not only to remove the volatile compounds from the gasoline but, also had to stimulate biodegradation, particularly of the nonvolatile components, by the (passive) infiltration of air (oxygen). The unsaturated zone of the soil consisted of fine sand to gravel. The groundwater level had to be lowered from 2 to 3 m below ground surface to enlarge the unsaturated zone and make the smear zone available for SVE. Figure 1 is a diagram of a cross-section of the site. On one side of the road there are several SVE wells (perforated 2 to 2.75 m below ground surface), and on the other side there are seven infiltration wells (passive). To prevent direct air infiltration at the extraction side of the road, a plastic liner was placed between the road and the sheet pile wall. The biological system applied for combined soil vapor and groundwater treatment was BIOPUR®.

Results and Discussion. Soil vapor was withdrawn at a rate of 60 Nm^3/hour at a vacuum of 100 mbar. The concentration of volatile hydrocarbons in the withdrawn soil vapor has decreased from 160 g/m^3 to 0.5 g/m^3. Figure 2 gives the cumulative amount of vaporized and biodegraded hydrocarbons as well as the sum of both. The vaporized amount takes into account both the airflow and concentration. The biodegraded amount is derived from the withdrawn mass flow of CO_2.

The CO_2 concentration in the withdrawn soil vapor amounted to approximately 0.5% (v/v) from week 0 to 37, increased to approximately 1% (v/v) in weeks 37 to 47, and gradually decreased to 0.25% (v/v) over weeks 47 to 118. The biological removal rate appeared to be approximately 7 mg C/kg/day.

The removal of the volatile hydrocarbons was in accordance with expectations. The degradation rate decreased in the final stages of the in situ remediation. To speed up the in situ bioremediation, it was decided to take additional measures (commencing week 78). These measures consisted of (heated) air injection (35 to 40°C), fluctuation of the groundwater level, and infiltration of nutrients (N/P).

Conclusions. In practice SVE proved to be successful as more than 4,000 kg of hydrocarbons were removed and the concentrations decreased from 10,000 mg/kg to <100 to 260 mg/kg within two years at Site 1. These results were obtained using supplementary measures, listed above. The remediation was stopped because it had been completely successful.

SITE 2: IN SITU SOIL
VAPOR EXTRACTION
COMBINED WITH AIR INJECTION

Characteristics. An underground tank, filled with diesel, leaked and caused more than 20,000 m^3 of soil to be contaminated. Due to the relatively low groundwater level (5 m below ground surface) and the soil

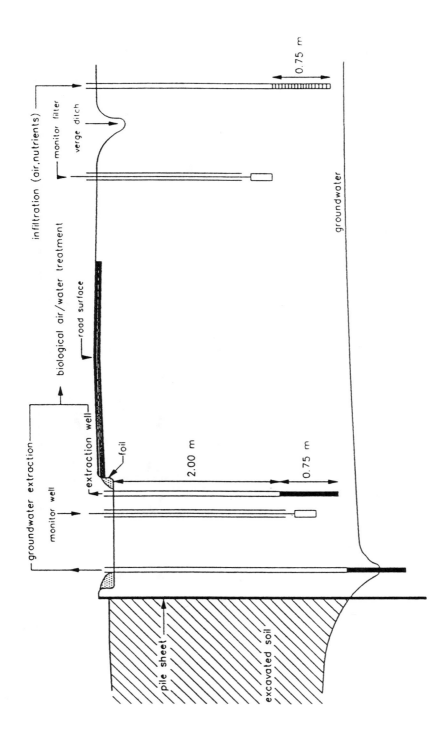

FIGURE 1. Cross-section of Site 1.

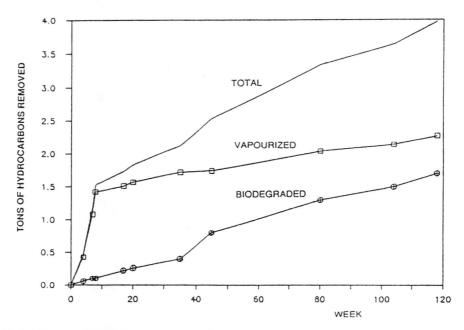

FIGURE 2. Cumulative amount of vaporized and biodegraded hydro-carbons.

profile (grain sand, hydraulic conductivity 300 m/d, the underground tank was situated in a clay layer above), more than 30,000 m³ of clean soil, situated on top of the contaminated soil layer, would have to have been excavated. An interesting alternative to the problem was to bio-remediate the soil using SVE coupled with air injection. The SVE will remove the volatile organic compounds and the air injection will saturate the groundwater with oxygen to enhance bioremediation in the saturated zone and to remediate the groundwater and soil. A feasibility study was performed on laboratory scale that soon showed the in situ approach to be feasible.

A test was conducted on pilot plant scale so that data could be obtained to design an appropriate remediation system. Figure 3 gives a cross-section of the site. An air injection well was placed 10 m below ground surface. Seven vapor extraction wells were installed to withdraw the soil vapor. Groundwater wells were fitted to monitor the concentrations of contaminants, nutrients, and oxygen in the aquifer.

Results and Discussion: Soil Vapor Extraction. The air was injected at a rate of 70 Nm³/hour and withdrawn at a rate of 180 Nm³/hour to inhibit

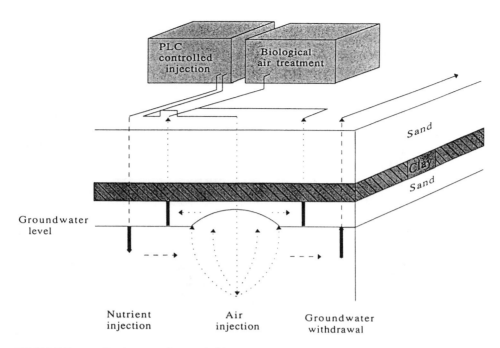

FIGURE 3. Cross-section of Site 2.

the contaminants from migrating. The concentration of hydrocarbons withdrawn from the soil has decreased from 1,000 mg/m^3 to 100 mg/m^3. After 4 months, 120 kg of hydrocarbons have been removed using SVE.

The amount of CO_2 present in the withdrawn air was also measured. The amount of CO_2 production cannot be ascribed to the site at the pilot plant alone but reflects the CO_2 production of the entire site. A constant concentration of 1.4% (v/v) CO_2 was found during the 4-month experiment. In one experiment, after 3 months CO_2 production was measured in the soil while no air was injected. CO_2 concentration rose up to 3.5% (v/v). From this experiment, a carbon consumption rate of 1.5 mg C/kg/day can be derived.

Results and Discussion: Air Injection. Air was injected at a rate of 70 Nm^3/hour at various pressures ranging from 0.5 to 3 bar. The influence of the air injection was measured by monitoring the groundwater level and the O_2 concentration in the groundwater, and by tracer experiments in water and air. Figure 4 outlines the oxygen level in the groundwater versus the distance from the air injection. The radius of influence is approximately 6 to 8 m in this configuration.

The rise in the groundwater level caused by air injection is as much as 90 cm. To check whether or not groundwater flow is created by this effect, tracer experiments were carried out by injecting a liquid tracer through the air injection system. Due to the rise of the groundwater level by air injection, a groundwater flow rate of 3 to 4 m/day was expected. However, the tracer was not detected in the groundwater monitoring wells, six meters from the injection well, in the forty days after the tracer was injected, indicating that air injection did not create groundwater flow.

Figure 5 shows the oxygen consumption when air injection has ceased. Two situations are outlined, one with nutrients and the other without. Nutrients (N and P) were added to the air injection well. Oxygen consumption appears to be 5 to 8 times higher when nutrients have been added to the soil.

Within four months, concentrations of nonvolatile hydrocarbons in the groundwater monitoring filters have decreased from 13,000 µg/L to less than 100 µg/L (detection limit) within the air injection radius of influence.

Conclusions. Air injection is an effective method of saturating large volumes of groundwater with oxygen. No forced groundwater flow has been observed due to the air injection. Due to the injection of nutrients (N and P) into the subsoil biological activity has increased by a factor of 8. Approximately 120 kg of hydrocarbons were removed within four months. A remedial action plan for the entire site, has been set up based on these results.

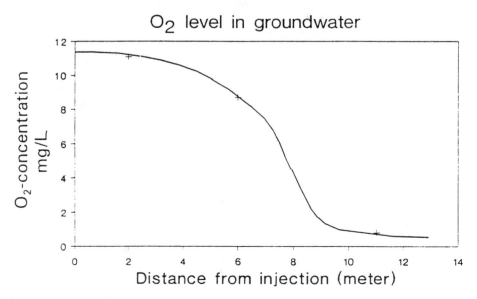

FIGURE 4. Air injection radius of influence.

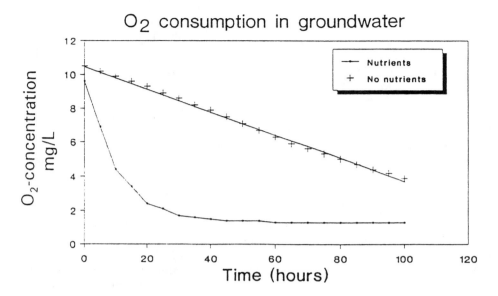

FIGURE 5. Oxygen consumption with and without nutrients.

CONCLUSIONS

In situ soil remediation using SVE combined with air injection has proved to be a successful technique in the treatment of soil contaminated with hydrocarbons. Its success is due to a combination of withdrawing the relatively volatile contaminants and stimulating biological activity in the subsoil to bioremediate the nonvolatile hydrocarbons. By monitoring the O_2 and CO_2 levels in the soil, data were obtained about the biological activity of the subsoil.

ACKNOWLEDGMENTS

The Site 2 project was financed by the Government House of Limburg and was realized by TAUW Infra Consult B.V. and Oosterhof Holman Milieutechniek B.V.

REFERENCES

Coffa, S., L.G.C.M. Urlings, and J.M.H. Vijgen. 1991. "Soil Vapor Extraction of Hydrocarbons In Situ and On Site Biological Treatment at a Gasoline

Station." NATO/Committee on Challenges of Modern Society (NATO/CCMS), International meeting Washington DC, November 1991.

Hinchee R. E., and R. N. Miller. 1990. "Bioreclamation of Hydrocarbons in the Unsaturated Zone." *Hazardous Waste Management Contaminated Sites and Industrial Risk Assessment*, pp. 641-650. W. Pillmann and K. Zirm (Eds.), Vienna, Austria.

Hutzel, N. J., B. E. Murphy, and S. Gierke. 1989. *State of Technology Review, Soil Vapor Extraction System.* EPA/600/2-89/024. U.S. Environmental Protection Agency.

Miller, R. N., R. E. Hinchee, C.C.M. Vogel, R. Ryan Dupont, and D. C. Downey. December 1991. "A Field Scale Investigation of Enhanced Petrol Hydrocarbon Biodegradation in the Vadose Zone at Tyndall AFB, Florida." *Proceedings NATO/CCMS*, Augers, France. pp. C108-C120.

Staps, J. J. 1990. *International Evaluation of In Situ Biorestoration of Contaminated Soil and Groundwater.* National Institute of Public Health and Environmental Protection, R.I.V.M. Report 738708006, The Netherlands.

Urlings, L.G.C.M., F. Spuij, S. Coffa, and H.B.R.J. van Vree. 1991. "Soil Vapor Extraction of Hydrocarbons - In Situ and On Site Biological Treatment." In R. E. Hinchee and R. F. Olfenbuttel (Eds.), *In Situ Bioreclamation.* Butterworth-Heinemann, Stoneham, MA.

AUTHOR LIST

D. P. Ahlfeld
University of Connecticut
Environmental Research Institute
Route 44, Longley Bldg.
Box U-210
Storrs, CT 06269 USA

E. J. Alesi
IEG Technologies Corporation
1833 D Crossbeam Drive
Charlotte, NC 28217 USA

G. K. Billings
Billings & Associates, Inc.
3816 Academy Parkway N-NE
Albuquerque, NM 87109 USA

J. F. Billings
Billings & Associates, Inc.
3816 Academy Parkway N-NE
Albuquerque, NM 87109 USA

J. W. Borthen
ECOVA Corporation
18640 NE 67th Ct.
Redmond, WA 98052 USA

G. Bott-Breuning
Dettinger Strasse 146
D-7312 Kirchheim/Teck GERMANY

J. Brauns
Institute for Soil Mechanics and Rock
    Mechanics
University of Karlsruhe
Richard-Willstätter-Allee
76 128 Karlsruhe GERMANY

R. A. Brown
Groundwater Technology, Inc.
310 Horizon Center Drive
Trenton, NJ 08691 USA

W. Buermann
Institute of Hydromechanics
University of Karlsruhe
Kaiserstrasse 12
D-7500 Karlsruhe 1 GERMANY

A. I. Cooley
Billings & Associates Inc.
3816 Academy Parkway N-NE
Albuquerque, NM 87109 USA

M. A. Dahmani
University of Connecticut
Environmental Research Institute
Route 44, Longley Bldg.
Box U-210
Storrs, CT 06269 USA

S. Diekmann
Institute of Hydromechanics
University of Karlsruhe
Kaiserstrasse 12
D-7500 Karlsruhe 1 GERMANY

M. Farrell
University of Connecticut
Environmental Research Institute
Route 44, Longley Bldg.
Box U-210
Storrs, CT 06269 USA

T. C. Hazen
Savannah River Technology Center
Westinghouse Savannah River
    Company
Aiken, SC 29808 USA

B. Herrling
Institute of Hydromechanics
University of Karlsruhe
Kaiserstrasse 12
D-7500 Karlsruhe 1 GERMANY

P. M. Hicks
Groundwater Technology, Inc.
310 Horizon Center Drive
Trenton, NJ 08691 USA

R. J. Hicks
Groundwater Technology, Inc.
310 Horizon Center Drive
Trenton, NJ 08691 USA

R. E. Hinchee
Battelle
505 King Avenue
Columbus, OH 43201-2693 USA

W. Ji
University of Connecticut
Environmental Research Institute
Route 44, Longley Bldg.
Box U-210
Storrs, CT 06269 USA

R. L. Johnson
Oregon Graduate Institute
19600 NW Von Neumann Drive
Beaverton, OR 97006 USA

F. Li
Vapex Environmental Technologies, Inc.
480 Neponset Street
Canton, MA 02021 USA

K. H. Lombard
Bechtel Savannah River, Inc.
Building 704-B
Savannah River Site
Aiken, SC 29808 USA

M. C. Marley
Vapex Environmental Technologies, Inc.
480 Neponset St.
Canton, MA 02021 USA

C.G.J.M. Pijls
Post Office Box 479
7400 Al Deventer
THE NETHERLANDS

F. Spuij
TAUW Infra Consult B.V.
P.O. Box 479
7400 AL Deventer
THE NETHERLANDS

J. Stamm
Institute of Hydromechanics
University of Karlsruhe
Kaiserstrasse 12
D-7500 Karlsruhe 1
GERMANY

L.G.C.M. Urlings
Post Office Box 479
7400 Al Deventer
THE NETHERLANDS

H.B.R.J. van Vree
TAUW Infra Consult B.V.
P.O. Box 479
7400 AL Deventer
THE NETHERLANDS

K. Wehrle
Institute for Soil Mechanics and Rock
     Mechanics
University of Karlsruhe
Richard-Willstätter-Allee
76 128 Karlsruhe
GERMANY

INDEX